4/08 m

2004

MAR

# Five Golden Rules

# Five Golden Rules

## Great Theories of 20th-Century Mathematics— and Why They Matter

John L. Casti

**John Wiley & Sons, Inc.**
New York ∎ Chichester ∎ Brisbane ∎
Toronto ∎ Singapore

To Ken Canfield—who asked for it

Library of Congress Cataloging-in-Publication Data

Casti, John L.
    Five golden rules : great theories of 20th-century mathematics—and why
they matter / John L. Casti.
        p.    cm.
    Includes index.
    ISBN 0-471-00261-5
    1. Mathematics—Popular works.   I. Title.   II. Title: 5  golden
rules
    QA93.C38   1996  510'.9'04—dc20                                94-44470

Printed in the United States of America
10 9 8 7 6 5 4 3 2 1

# CREDITS

Grateful acknowledgment is made to the following for permission to reproduce material used in creating the figures in this book. Every reasonable effort has been made to contact the copyright holders of material used here. Omissions brought to our attention will be corrected in future editions.

Figure 1.2, Reprinted with permission of The Free Press, a Division of Simon & Schuster from *Game Theory and Politics,* by Steven J. Brams. Copyright © 1975 by The Free Press.

Figure 1.3, Venttsel, E., *An Introduction to the Theory of Games,* DC Heath, 1963.

Figure 1.4, Thomas, L.C., *Game Theory,* Ellis Horwood, UK, (ND). Reprinted with permission of the author.

Figure 2.1, *Science News.*

Figures 2.2 and 2.5, Stewart, I., *Concepts of Modern Mathematics,* Penguin Books, Ltd., 1975. Reprinted with permission of the author.

Figures 2.3, 2.4, 2.9, Jacobs, K., *Invitation to Mathematics,* Copyright © 1992 by Princeton University Press. Reproduced by permission of Princeton University Press.

Figure 2.7, Los., J., *Computing Equilibria: How and Why,* Polish Scientific Publishers, 1975.

Figure 2.11, Scarf, H., Fixed-point theorems and economic analysis. *American Scientist,* May–June 1983. Reprinted with permission of Sigma Xi.

Figure 2.13 and 2.14, Smart, D.R., *Fixed-Point Theorems,* Copyright © 1974 by Cambridge University Press. Reprinted with the permission of Cambridge University Press.

Figure 2.15, Fred S. Roberts, *Discrete Mathematical Models,* ©1976, p. 267. Reprinted by permission of Prentice Hall, Englewood Cliffs, New Jersey.

Figure 3.1, Kaye, B., *Chaos and Complexity,* Fig. 1.3. New York: VCH Publishers. Reprinted with Permission by VCH Publishers © 1994.

Figures 3.2 and 3.3, Callahan, J., Singularities of plane maps. *Am Math Monthly,* March 1974.

Figures 3.4, 3.5, and 3.6, Poston, T. and I. Stewart, *Catastrophe Theory and its Applications*, Copyright © 1978 by Pitman Pubishing. Reprinted with permission of the Longman Group, Ltd.

Figure 3.8, Poston, T. and I. Stewart, *Taylor Expansions and Catastrophes,* Copyright © 1976 by Pitman Pubishing. Reprinted with permission of the Longman Group, Ltd.

Figure 3.9, Arnold, V., *Theory of Singularities and Its Application,* Copyright © 1981 by Cambridge University Press. Reprinted with permission of Cambridge University Press.

Figure 3.10, Francis, G., *A Topological Picture Book,* Fig. 2. New York: Springer-Verlag, Inc., 1987. Reproduced with permission of Springer-Verlag.

Figure 3.11, Gilmore, R., *Catastrophe Theory for Scientists and Engineers,* Fig. 5.3. New York: John Wiley & Sons, 1981. Reproduced by permission of the author.

# Contents

# Preface

The lifeblood sustaining any field of intellectual endeavor is the infusion of a steady stream of important, unsolved (but in principle solvable) problems. Projective geometry, for example, once a flourishing corner of the mathematical forest, is nowadays about as dead as the dodo bird for the simple reason that the wellspring of good problems ran dry about a hundred years ago. On the other hand, the currently fashionable rage for chaos was totally unknown to all but a few far-sighted adventurers and connoisseurs of the mathematically arcane until the rather recent work of Lorenz, Smale, Feigenbaum, Yorke, May, Rössler, and many others stimulated the outpouring of problems that sustain today's chaologists, their students, and camp followers. These examples illustrate clearly George Polya's well-known dictum that "Mathematics is the art of problem solving." But unlike scientists in other disciplines, mathematicians have a special word for the solution to one of their problems—they call it a *theorem.*

Mathematics is about theorems: how to find them; how to prove them; how to generalize them; how to use them; how to understand them. *Five Golden Rules* is intended to tell the general reader about mathematics by showcasing five of the finest achievements of the mathematician's art in this century. The overall plan of the book is to look at a few of the biggest problems mathematics has solved, how they've been solved and, most importantly, why the solutions matter—and not just to mathematicians. Thus, the goal of *Five Golden Rules* is to enlighten, entertain, and educate by example, rather than by precept.

Stanislaw Ulam once estimated that mathematicians publish more than 200,000 theorems every year. The overwhelming majority of these are completely ignored, and only a tiny fraction come to be understood and believed by any sizable group of mathematicians. Given the fact that mathematics has been practiced on this planet for several millennia, at first sight it seems a daunting prospect to try to single out the "greatest" theorems even of this century from a list that by now must number well into the millions. But the task can soon be cut down to size by the imposition of a small number of conditions, or "filters," separating the great theorems from the pretenders. To pinpoint the five jewels highlighted in this book, here are the criteria I employed:

- *Significance:* Did the theorem break a major logjam in the development of mathematics? Or did the result lead to the establishment of new fields of mathematical enquiry? *Example:* Morse's Theorem, which sparked off the development of singularity theory.

- *Beauty and Scope:* Is the theorem intrinsically "beautiful," in just the same sense that a poem or a painting is beautiful? Does it summarizes compactly a large body of knowledge? And does the theorem shed light on questions over a broad range of areas *inside* mathematics? *Example:* Brouwer's Fixed-Point Theorem, which enables us to establish the existence of solutions to equations under very general mathematical conditions in a wide variety of settings.

- *Applications:* Does the theorem find important applications *outside* mathematics? Do the mathematical structures whose existence the theorem underwrites provide the basis for a more complete understanding of the world of nature and/or humankind? *Example:* The Minimax Theorem, which forms the cornerstone of much of the mathematical work in economics and elsewhere on what it means to say the actions of decisionmakers are "rational."

- *Proof Method:* Did the proof of the theorem require the use of new techniques of logic or modes of reasoning? Could these methods be used to make major inroads on other important problems? *Example:* The Halting Theorem, whose proof focused attention on the idea of using an algorithm to establish mathematical truths.

- *Philosophical Implications:* Does the theorem tell us something important about human beings that we didn't know before? Do the theorem's conclusions impose major restrictions or, conversely, open up new opportunities for us to obtain deeper insights into what it is we can know about the universe and about ourselves? *Example:* Gödel's Incompleteness Theorem, which imposes limitations upon the ability of the human mind to formalize real-world truths.

In order to qualify for inclusion on our roll call of honor, a theorem would have to score high in most, if not all, of these categories. It doesn't take too much imagination to see that employing these filters quickly whittles down Ulam's universe of millions of theorems to manageable size.

But great theorems do not stand in isolation; they lead to great theories. As indicated above, an important part of the significance of a theorem lies in the theories it contributes either to creating or in some way to nourishing. And for this reason, our focus here is at least as much on great *theories* of twentieth-century mathematics as it is on the great theorems themselves.

A quick glance at the book's contents might lead the reader to ask, Why are the theorems considered here so old? The most recent entry on the list of the Big Five is the Simplex Method, which dates back to around 1947, while the earliest is Brouwer's Fixed-Point Theorem, which was published in 1910. If it's modern, that is, twentieth-century, mathematics we're after, why is there nothing from the work of the past 50 years? This is especially puzzling when by common consensus more significant mathematical work has been done in the latter half of this century than in all previous centuries combined.

This is a pretty reasonable question, so it deserves a carefully considered reply. Basically, the answer resides in the fact that it's really the great *theories* we're after, not the great *theorems*. And great theories in mathematics are like great poems, great paintings, or great literature: it takes time for them to mature and be recognized as being "great." This brings to mind a remark made by Michael Faraday to a British prime minister who was visiting his workshop. When Faraday described his latest discovery in electricity, the distinguished gentleman asked, "What good is it?" Faraday replied, "What good is a newborn baby? You have to wait for it to grow up." And so it is with great theorems, as well. Generally speaking, it seems to take at least a generation or two for a great theorem to "grow up," that is, to be recognized as the seed from which a great theory has subsequently grown. So, what we see as a great theory today almost necessarily had its origin in results dating from the pre–World War II era. And I have no doubt that a similar book written ten years from now will focus on theorems of the 1960s and 1970s, which only now are starting to crystallize in the form of still more great theories. Let me note that sometimes a great theory requires advances in technology, too. For instance, I doubt seriously that two of the great theories treated in this book—optimization theory and the theory of computation—would have appeared in any such volume had it not been for the major advances that have taken place in computing technology over the past few decades.

When the idea for this book first struck me, I queried a number of friends and mathematical scholars as to what they would include in a volume addressing the great theorems and theories of twentieth-century mathematics. Someday I'd like to publish that list, which unfortunately is a bit too long to comfortably include here. But when I had made the final choices, someone asked me why the book was so "impure"; why were all the theories (with the possible exception of topology) in areas that some euphemistically (or pejoratively!) call "applied mathematics" (a term, incidentally, that I abhor). Why is there nothing here that might be termed "pure" mathematics? The reasons are twofold: (a) we're all prisoners of our tastes and background, and mathematically speaking, mine lean in the applied directions, and (b) I wanted the book to focus on why ordinary people (that is, nonacademics) should care about mathematics as a factor in their daily lives, a goal that again biased the material to the applied side of the house. I would certainly look forward to a similar book by someone with leanings different from mine, dealing perhaps with great modern results (and theories) such as the Atiyah-Singer Index Theorem (partial differential equations), The Classification Theorem for Finite, Simple Groups (group theory) and the Hahn-Banach Theorem (functional analysis). But I don't think I will be the person to write that book.

Since I've claimed that the book is for those who want to know about mathematics and why it matters, I'm at least implicitly saying that this is a book aimed at the nonmathematician. This fact requires some "deconstruction." At the outset, let me say that to write about mathematics using no mathematics *at all* is, in my opinion, a copout, doing a disservice both to an intelligent reader and to the field of mathematics itself. To write such "baby talk" about mathematics requires either treating only topics that lend themselves to drawing pictures, like geometry and fractals, or discussing puzzles involving the properties of numbers, simple probability theory or elementary logic that as often as not miss the excitement—and the content—of where the real mathematical action is taking place. So I have chosen a different, less journalistic—and far more dangerous—route. The path this book has taken was dictated long ago by Einstein, when he stated, "A theory should be as simple as possible— but no simpler." So allow me to try to translate this semicryptic remark into a statement about the reader at whom I am aiming this book.

The target reader for the material presented here will have a background in mathematics that I like to call *sophisticated.* This does *not* mean that he or she actually knows any mathematical techniques or procedures—there is very little by way of technical mathematics in this book (none, actually), but there is a *lot* by way of mathematical concepts, ideas, and chains of reasoning. Moreover, I do not believe in the well-chronicled statement by Stephen Hawking's editor to the effect that every equation in a book cuts its sales in half. My ideal reader won't believe it, either. The odd equation will turn up from time to time in the pages to follow, as will an occasional Greek symbol and even a graph or two. But a reader who cares about learning what mathematicians have achieved—and why it matters—won't be deterred in the least by such formalities. He or she will swat these low-level barriers aside as if they were nothing more than pesky mosquitos. So the book is for anyone who's not afraid to confront real mathematical *ideas*—head-on. Just about anyone who's had a course in high-school algebra or geometry, and who retains at least a modicum of enthusiasm for mathematical ideas fits into this category—even if the details of that long-ago course have faded from memory. It's the ideas and one's willingness to confront them that counts, not the technical details.

What about those who do have a more detailed background in mathematics? If comments on the draft versions of the book are anything to judge by, even many professional mathematicians will find material in the book to interest them. Of course, what they will *not* find is the kind of rigor and detailed proofs that one expects from a mathematical textbook or research monograph. This is not a textbook for "wannabe" mathematicians (although it has worked well as a text in undergraduate liberal arts courses of the "mathematics-for-poets" type). And it certainly is not a research monograph. It's pure exposition. But for those who care to dig deeper into the details of the arguments that are only sketched in broad outline here, I have liberally sprinkled the book's bibliography section with a number of more detailed accounts of each topic at various levels of mathematical difficulty.

As always in putting together a work such as this, information and encouragement from many sources was invaluable. So let me close this already overly long preface by performing that most pleasurable of all tasks associated with the writing of any book, namely, the tipping of my hat to those friends and colleagues who have given so generously of their time on behalf of the book. Ian Stewart, Phil Davis, Don Saari, Martin Shubik, Atlee Jackson, and Greg Chaitin offered their thoughts and wise counsel on both the content and style of the book. In this same regard, I wish to single out for special honors my former teacher, friend, and now colleague, Tom Kyner. His careful reading and comments on virtually every line of the original manuscript materially improved the final version, as well as saved me from several flat-out technical faux pas and other inanities and infelicities. For TeX typesetting consultations, it's always a pleasure to acknowledge Michael Vulis and Berthold Horn. And Jennifer Ballentine of Professional Book Center was a font of wisdom (no pun intended) and advice when it came to matters of book design.

Finally, accolades *magna cum laude* to the book's editor, Emily Loose, who has been a constant source of encouragement and eagle-eyed editing, both of which contributed mightily to a far better final product than I had any right to expect. My thanks to all of the above and my absolutions, as well, for any and all errors that managed somehow to creep into the final product. These, I'm sorry to say, remain solely my responsibility.

JLC
Santa Fe, New Mexico

# CHAPTER

# 1

# The Minimax Theorem

*Game Theory*

# Deadly Games

In everyday conversation, a "game" is often thought of as a mere pastime for schoolchildren, a way to spend their day avoiding homework and piano lessons, perhaps playing instead something like blindman's bluff, tag, or hide-and-seek. But to many adults, the term also conjures up images of ascetic chess players hunched over boards in smoke-filled cafes or captains of industry in equally smoke-filled corporate boardrooms, all desperately seeking strategies that will give them an advantage over their opponent(s). These latter situations, in which the outcome of the game is determined by the strategies employed by the players, form the starting point of what we now term the mathematical *theory of games*. And the essential ingredient making game theory a "theory" rather than a collection of heuristics, rules of thumb, anecdotal evidence, and old wives' tales is the notion of a minimax point, a set of optimal strategies for all players in the game. Let's begin with a very real-world example illustrating the general idea.

In early 1943, the northern half of the island of New Guinea was controlled by the Japanese, while the Allies controlled the southern half. Intelligence reports indicated that the Japanese were assembling a convoy to reinforce their troops on the island. The convoy could take one of two different routes: (1) north of New Britain, where rain and bad visibility were predicted, or (2) south, where the weather was expected to be fair. It was estimated that the trip would take 3 days on either route.

Upon receiving these intelligence estimates, the Supreme Allied Commander, General Douglas MacArthur, ordered General George C. Kenney, commander of the Allied Air Forces in the Southwest Pacific Area, to inflict maximum possible damage on the Japanese convoy. Kenney had the choice of sending the bulk of his reconnaissance aircraft on either the southern or the northern route. His objective was to maximize the expected number of days the convoy could be bombed, so he wanted to have his aircraft find the convoy as quickly as possible. Consequently, Kenney had two choices: (1) use most of his aircraft to search the northern route, or (2) focus his search in the south. The payoff would then be measured by the expected number of days Kenney would have at his disposal to bomb the convoy. The overall situation facing the two commanders can be represented in the "game tree" of Figure 1.1, which summarizes what came to be termed the Battle of the Bismarck Sea.

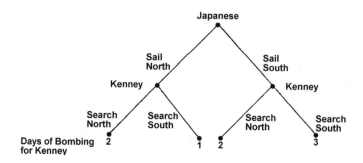

**Figure 1.1** Game tree for the Battle of the Bismarck Sea.

(Technically, this kind of game tree is what game theorists call the *extensive form* of the game.)

The diagram should be read in the following way: Starting at the top node, the Japanese commander can choose either the left branch (Sail North) or the right branch (Sail South). Each of these branches leads to a node labeled "Kenney," indicating that these nodes are decision points for General Kenney. The choices for Kenney are now to take the left branch (Search North), or to select the right branch (Search South). After the two commanders have made their choices, the tree "bottoms out" at one of the numbers listed below each of the termination nodes. This number is the days of bombing intelligence estimates claim are available to Kenney if the decisions of the two commanders led to that particular endpoint. In reality, of course, the commanders did not make their choices in the sequential fashion suggested by the diagram. Rather, each chose his course of action independently, without knowledge of what the other was going to do.

It's clear that in making their decisions, General Kenney and the Japanese commander have diametrically opposed interests: What's good for General Kenney is bad for the Japanese commander, and vice versa. Thus, we measure the payoff to the Allies as the number of days of bombing, while we count the "reward" to the Japanese as the negative of this number. So what one side wins, the other side loses. This is an example of what's called a "zero-sum situation," since the payoffs to the two commanders add up to zero.

A more compact way of expressing the overall situation is to use what's termed a *payoff matrix*, which defines the *normal form* of the game. It is shown below for the Battle of the Bismarck Sea. The rows of the matrix are labeled with the choices available to General Kenney, while the columns represent the alternatives at the disposal of the Japanese commander. By convention, the entries in the matrix are taken to be the payoff to the "row player," in this case General Kenney and the Allies. Since the interests of the Allies and the Japanese are directly opposed, the payoffs to the "column player," the Japanese, are simply the negatives of these numbers. By reference to the game tree shown in Figure 1.1, we see that these payoffs are the same as the numbers that terminate each path from the top of the tree to the bottom. Given these possible courses of action and payoff structure, the problem is to determine what a *rational* commander should do.

|  |  | Japanese | |
|---|---|---|---|
|  |  | **Sail North** | **Sail South** |
| Allies | **Search North** | 2 days | 2 days |
|  | **Search South** | 1 day | 3 days |

Payoff matrix for the Battle of the Bismarck Sea

We have assumed that Kenney wants to maximize the number of days that he can bomb the Japanese. If he searches north, Kenney is guaranteed 2 days of bombing regardless of the direction the Japanese decide to sail. But if he searches south, then Kenney may only get a single day of bombing if the Japanese fleet sails north. But if they sail south, he will get 3 days of bombing. So to avoid having any regrets once he finds out what the Japanese have decided to do, Kenney should elect to *maximize* the minimum number of days he can bomb the Japanese fleet. This implies that he should search north. By a similar chain of reasoning, the regret-free choice of the Japanese commander is to sail north, since that is the decision that *minimizes* the maximum number of days that his fleet is exposed to the Allied bombers.

Generalizing from this elementary, commonsense analysis of the situation facing General Kenney and the Japanese commander, we conclude that the rational decisionmaker looks for a course of action that will give him or her the best possible payoff in a worst-case situation, that is, the best payoff assuming that one's opponent makes his or her best counter move. Clearly, this results in the players each taking what might be called a "risk-averse" decision: one that forsakes possible gains in order to avoid incurring unnecessary losses. Now let's see how these risk-averse, rational courses of action can be obtained directly from a game's payoff matrix.

Recalling the payoff matrix for the Battle of the Bismarck Sea, the row player (the Allies) wants to maximize its minimum payoff. So we write down the minimum entry in each of the rows of the matrix, which we recall represent the payoffs for actions available to General Kenney. Similarly, the column player (the Japanese commander) wants to minimize the maximum number of bombing days they are exposed to. Consequently, we write down the maximum of each column of the matrix. After this exercise, we arrive at the array shown below:

|  |  | Japanese | | |
| --- | --- | --- | --- | --- |
|  |  | **Sail North** | **Sail South** | *Row Min* |
| Allies | **Search North** | 2 days | 2 days | **2** |
|  | **Search South** | 1 day | 3 days | 1 |
|  | *Column Max* | **2** | 3 | |

By our risk-averse notion of what constitutes a rational choice, the Allies want to find the largest of the row minima, while the Japanese seek the smallest of the column maxima. Lo and behold, these two numbers (set in boldface in the array) coincide at the pair of decisions: Search North for the Allies and Sail North for the Japanese. Such a combination of actions, at which the maximum of the row minima (the "maxmin") equals the minimum of the column maxima (the "minmax"), is called an *equilibrium point* of the game. This is because by choosing these actions the two players guarantee themselves a certain minimal payoff—regardless of what their opponent happens to do. Thus, neither player is motivated to unilaterally depart from his or her equilibrium decision. Furthermore, neither player will have cause to regret this decision once the opponent's choice is known, because both notice that given their adversary's choice, he or she would have done worse by choosing differently. In other words, the equilibrium solution is "bulletproof," or stable, in the sense that either player can announce his or her choice in advance to the opponent, secure in the knowledge that the opponent cannot make use of this information to get a better payoff.

An equilibrium decision point of this sort is often termed a *saddle point*. This nomenclature arises by thinking of the payoff $z$ for a play of the game as being given by a number that depends on the choice made by the row player, call it $x$, and the decision of the column player, which we shall label $y$. In other words, the payoff is a real-valued function of $x$ and $y$, which we will call $a(x, y)$. If $x$, $y$ and $a(x, y)$ are real numbers, then the function $a(x, y)$ can be represented geometrically by a surface in three-dimensional space, as shown in Figure 1.2. Regarding this surface as an altitude map, in which the choices $x$ and $y$ are lines of constant latitude and longitude, respectively, the payoff $z = a(x, y)$ is then simply the altitude where these two lines intersect. The row player wants to get to the highest peak on this surface, while the column player wants to get to the lowest valley. So if there is a point like $S$ in the figure that is simultaneously the highest point in the $x$ direction and the lowest in the $y$ direction, then we have something that looks like the saddle-shaped surface shown in the figure. It is the geometry of this situation that accounts for the terminology "saddle point." Such a point is also often called a *minimax point*.

In the Battle of the Bismarck Sea, the pair of choices $x$ = Search North and $y$ = Sail North is just such a saddle point. Interestingly, in the actual Battle of the Bismarck Sea, the two commanders did indeed take these decisions (which resulted in a crushing defeat for the Japanese).

What's important about a saddle point is that it represents a decision by the two players that neither can improve upon by unilaterally

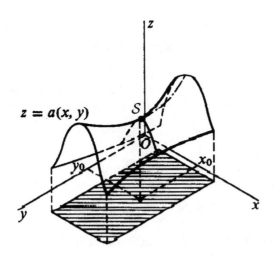

**Figure 1.2** A game-theoretic saddle point.

7

departing from it. In short, either player can announce such a choice *in advance* to the other player and suffer no penalty by doing so. Consequently, the best choice for each player is at the saddle point, which is called a "solution" to the game in *pure strategies*. This is because regardless of the number of times the game is played, the optimal choice for each player is to always take his or her saddle-point decision.

But we have seen geometrically that the saddle point is at the same time the highest point on the payoff surface in one direction and the lowest in the other direction. Put in algebraic terms using the payoff matrix, the saddle point is where the largest of the row minima coincides with the smallest of the column maxima. But it's certainly easy to imagine payoff matrices and geometric surfaces for which no such saddle point exists. In such cases, there is no obvious way for a player to act that cannot be exploited by an opponent who happens to gain advance knowledge of what that player is going to do. But since there always exists the possibility that information about one's intentions might be "leaked" to an adversary, how should a rational player proceed under such circumstances? This is the central question upon which all of the mathematical theory of games rests. But before answering it, let's pause to formalize what we have learned so far about the theory of games from this simple example of the Battle of the Bismarck Sea.

## Games of Strategy

The view of the Battle of the Bismarck Sea given above, while a wildly oversimplified version of the real situation, still illustrates perfectly what's involved in what mathematicians think of as a "game of strategy." Reduced to its basics, a game of strategy consists of three interrelated components:

- *Players:* A game involves a minimum of two players who have differing interests.
- *Actions:* At each stage of play the players choose their course of action from a set of possible decisions, which are not usually the same for each player. Often, the players must make their decisions without knowledge of the actions chosen by the other players. We will also assume that each player has only a finite number of possible decisions, although some interesting and important games

8

make use of decision sets that are infinite (for example, an interval of real numbers).

- *Payoffs:* After the decisions have been made, each player receives a certain payoff measured in a common unit for all players.

For the sake of terminology, let us agree that a rule for choosing an action is termed a *strategy*. If the rule says to always take the same action, it's called a *pure strategy;* otherwise, the strategy is called *mixed*. A *solution* to a game is simply a strategy for each player that gives each of them the best possible payoff, in the sense of being a regret-free choice.

Further elaboration of these conditions leads to a multitude of different types of games, some of which will be discussed below. For example, if the game involves a sequence of choices by each player, we obtain what's called a *multistage* or *iterated* game. Of special concern for the moment are games like the Battle of the Bismarck Sea, in which there are only two players whose interests are directly opposed. We term such situations *two-person, zero-sum games,* since the payoff to the maximizing player is exactly the negative of what's received by the minimizing player. Consequently, the total payoff to both players sums to zero. Let's talk about this class of games in a bit more detail.

## Two-Person, Zero-Sum Games

When we contemplate the games of everyday life, we usually think of things like the battle of the sexes, the game of business, war games, or even something like poker or Monopoly. These games all seem far more complicated than the kind of strategic games just outlined. And one of their complicating features is that they tend to involve many players. So it might seem little more than an academic exercise to devote any appreciable amount of attention to games having only two players, at least if the goal is to have our theory make contact with reality. Nevertheless, there are reasons—practical and mathematical.

First of all, two-person games have the overwhelming attraction that they can be solved completely when the interests of the players are in direct opposition (the zero-sum case). This means that we can calculate exactly what each player's optimal strategy should be for this class of games. Such solutions give us a firm point of departure for the study of multiperson games, as well as providing a benchmark against which to compare approximate results in more complicated situations. Moreover, a number of situations in real life that initially appear as multiperson games can be reduced to the two-player case. For example, political conflicts over things like budget bills in the U.S. Senate look initially like 100-player games, with each senator playing against 99 opponents. However, it doesn't take a very detailed reading of the daily newspaper to see that these games often reduce to the two-player case: liberals versus conservatives, Democrats against Republicans, or business facing off against labor, for example. This kind of conflation of players is often a good approximation to the real situation, and allows us to bring to bear all the mathematical machinery of the two-player case.

But the idea of pure competition inherent in the zero-sum, two-person game is often too big an assumption to swallow. Moreover, real life abounds in situations in which the players' interests are only partially opposed. But even for these so-called *cooperative, multiperson* games, we still have a wealth of both empirical and theoretical knowledge available in the two-player case. For example, the famed Prisoner's Dilemma Game that we will consider in detail later in the chapter has led to a host of insights into how cooperative behavior can evolve in populations that initially consist solely of competitors, such as supermarkets competing for customers, politicians seeking votes, or animals competing for food. At first glance, this kind of situation appears to be one involving a large number of players. Nevertheless, it turns out that many important features of the overall situation facing competitors in such a game can be captured by considering only pairwise interactions, since it's seldom the case that three or more competitors simultaneously face off against each other. This, in turn, leads to a sequence of two-person games.

So with these ideas in mind, we return to the basic quandary: How should a rational player proceed in a zero-sum, two-person game that has no saddle point? Let's begin by considering one such game, again a rather oversimplified military situation that arose during the American Revolutionary War.

## The Concord Arsenal Game

The British have decided to attack the American arsenal at Concord. While American intelligence has turned up this important piece of information, the Americans do not know which way the British will come:

10

by land or by sea. Unfortunately, the American force is too small to defend both routes, so the American high command must choose one or the other.

In actual point of fact, the British are low on ammunition. So if the two forces meet, the British will retreat. But if the forces do not meet, the British will take the arsenal and gain control of a considerable amount of ammunition. If that happens, both sides have to decide what action to take with respect to the British return from Concord. The Americans can choose either to lay an ambush on the known path of return, or to move in and immediately attack the British at the arsenal itself. By the same token, the British can either leave the arsenal immediately by day or wait and withdraw by night. These various possibilities lead to four different courses of action that can be taken by each side, labeled A-I, A-II, ... , B-III, B-IV, respectively, the As denoting American choices and the Bs the alternatives for the British:

A-1 = defend by land, then ambush,    B-I = go by sea, then leave immediately,
A-II = defend by land, then attack,    B-II = go by sea, then wait for night,
A-III = defend by sea, then ambush,    B-III = go by land, then leave immediately,
A-IV = defend by sea, then attack,    B-IV = go by land, then wait for night.

Now let's consider the payoffs. The first point to note is that the interests of the two forces are directly opposed. Thus, the Concord Arsenal Game is a two-player, zero-sum game—just like the Battle of the Bismarck Sea. So we arbitrarily agree to measure payoffs as being to the Americans, with the British receiving the negative of this amount in each case.

Since the British are low on ammunition, the advantage resides with the Americans if the two forces meet. This direct encounter happens if the Americans take actions A-I or A-II and the British respond with B-III or B-IV. It also occurs if the Americans choose A-III or A-IV and the British choose B-I or B-II. Any of these four combinations yields the best possible outcome for the Americans, so we give them a payoff of, say, 2 points for each of these pairs of choices. For the remaining entries in the matrix, we argue as follows. If the British encounter an ambush by day, they will be wiped out—which is also just fine with the Americans. This happens with the pairs of actions (A-I, B-I) or (A-III, B-III). So again score 2 points for the Americans in these cases. But if the British meet the ambush by night, they can get through, yielding a payoff of

0 points for the Americans. These are the pairs (A-I, B-II) and (A-III, B-IV). Now if the Americans attack the arsenal and the British have already left, the choices (A-II, B-I) and (A-IV, B-III), the Americans also score 0 points. But if the Americans attack and find the British waiting for a night withdrawal, both sides suffer heavy losses and the Americans receive only 1 point. These decisions correspond to the pairs (A-II, B-II) and (A-IV, B-IV). Putting all this information together, we can fill in the relevant entries in the payoff matrix as follows:

|  | British | | | |
|  | B-I<br>Sea<br>Immediately | B-II<br>Sea<br>at Night | B-III<br>Land<br>Immediately | B-IV<br>Land<br>at Night |
| Americans | | | | |
| **A-I Land Ambush** | 2 | 0 | 2 | 2 |
| **A-II Land Attack** | 0 | 1 | 2 | 2 |
| **A-III Sea Immediately** | 2 | 2 | 2 | 0 |
| **A-IV Sea Attack** | 2 | 2 | 0 | 1 |

A cursory inspection of the payoff matrix turns up the fact that the minimum of each row is 0; hence, the maximum of these minima is itself 0. On the other hand, the largest value in each column is 2, so the minimum of these maxima is 2. The maxmin and the minmax points do *not* coincide; there is no saddle point for this game. As a result, neither the American nor the British commander has a clear-cut, "bulletproof" course of action that he can take—and even announce in advance to his opponent—without running the risk of having serious regrets once his adversary's action becomes known. In short, the Concord Arsenal Game has no solution in pure strategies, since a premature disclosure by one of the commanders of a particular course of action would lead to his losing points that he would otherwise not have had to give up. For instance, in the absence of advance information on the British plans, either of the actions A-I or A-III looks good to the Americans since each yields the maximum payoff of 2 points for all but one of the British choices. But if the American commander knows in advance (perhaps through a report from a spy like Nathan Hale) that the British will go by sea and then wait and leave at night (action B-II), he would certainly avoid action A-I since it gives him the minimal payoff (and maximum regret) against that particular choice by the British. And, in fact, action A-I is the worst possible choice for the American commander if the British happen to tip their hand by announcing B-II in advance. So here we have an example of a game with no solution in pure strategies; there is no one best action for each side to take regardless of what the other side chooses to do. In short, there is no saddle point.

The situation we find ourselves in with such a game is analogous to that faced by workers in the early days of the theory of equations when trying to solve the equation $x^2 + 1 = 0$, which has no solution in real numbers. The historical resolution of this difficulty was to invent a new notion of number, namely, the complex numbers, in which everything gets set right again. In game theory, too, the way out of what we might term the "no-saddle-point dilemma" is to invent a new notion of what we mean by rational behavior.

We have seen that if the game has a saddle point each side can announce their action beforehand without in any way jeopardizing their best possible payoffs. So advance knowledge of what your opponent is going to do cannot help increase your take in this type of game—assuming, of course, that both you and your opponent act rationally. But without a saddle point all bets are off, since we have then lost our notion of what it means to be "rational." It turns out that the way to recapture this kind of "immunity from disclosure" is to act in such a way that you don't even know yourself what you're going to do until you do it! While at first sight this might sound paradoxical, especially in regard to what we normally think of as constituting "rational behavior," in the next section we show how this "no-knowledge" dictum can be translated into a workable generalization of the idea of an optimal strategy, a generalization that leads moreover to a resolution of the no-saddle-point dilemma.

## Keep 'Em Guessing

In his song "Bob Dylan's 115th Dream," Bob Dylan talks about having escaped from jail, and then having to decide whether to return to jail to help his friends or leave them to their fate and sail away on a ship. In the song, he lets the flip of a coin decide between the two possibilities. Dylan's decision-making procedure serves admirably to illustrate the principle underlying how to choose a course of action in a game without a saddle point.

13

In such games, it's crucial for the players to keep their choice secret from their opponent for the reasons outlined above. So what better way to keep your decision secret than to select it at random! This way, even the players themselves won't know what their actions will be until the choices are actually dictated by some randomizing mechanism, such as the spin of a roulette wheel or, perhaps, the flip of Bob Dylan's coin. Of course, it will generally be the case that some alternatives look more attractive than others. So not all possible actions will be given equal weight; what's needed to make a choice, then, is a biased coin or, perhaps, some weighted dice. Thus, we now regard a player's strategy as being given by a set of probabilities specifying the likelihood that any particular action is selected by the randomizing mechanism.

Of course, if the actions are selected randomly it no longer makes sense to speak of a definite payoff. Rather, the players will now measure their return for employing a certain strategy as the *expected* payoff obtained by using that strategy, where the expectation is computed by taking into account the relative likelihoods of taking the various actions available. To firmly fix these notions in mind, here's a familiar childhood game that illustrates optimal play in a game having no saddle point.

### The Rock-Scissors-Paper Game

This is a game for two players, call them Maximilian and Minerva (or just Max and Min, for short), who have to simultaneously select among three alternatives: Rock, Paper, or Scissors. The outcome of such a confrontation is given by the following rules:

1. Paper covers Rock.
2. Rock smashes Scissors.
3. Scissors cut Paper.

Assuming there is a payoff of 1 unit to the winner, the payoff matrix for the Rock-Paper-Scissors Game is given in the following array.

|       |       | Min |    |    |
|-------|-------|-----|----|----|
|       |       | P   | S  | R  |
|       | P     | 0   | -1 | 1  |
| Max   | S     | 1   | 0  | -1 |
|       | R     | -1  | 1  | 0  |

The first thing we note is that Max's gain is Min's loss and vice versa. So this is a zero-sum game. The second point is that the maximum of the row minima (−1) does not equal the minimum of the column maxima (+1); hence, the game has no saddle point. This means that if Max "peeks" he can gain an advantage over Min, since he can play S against P, R against S, and P against R, and thus always prevail. Consequently, if both contestants play honestly, then each has to independently choose from the three possibilities in a random manner. And since the structure of the payoff matrix is completely unbiased with respect to the three choices, there is no a priori reason to favor one choice over the others. So both Max and Min will choose from among the three options with equal frequency, 1/3 for each. As a result, each of the nine possible combinations PP, PS, ... , RR will occur with likelihood 1/9. On the average, then, the payoff to both Max and Min will be

$$\frac{1}{9}(0 - 1 + 1 + 1 + 0 - 1 - 1 + 1 + 0) = 0. \qquad (\dagger)$$

Now suppose Min continues to select among P, S and R with likelihood 1/3 for each, but that Max changes his frequencies to, say, P = 1/3, S = 1/2, and R = 1/6. Now the average, or expected, outcome for Max (and Min, too, as it turns out) is

$$\left(\frac{1}{3} \times \frac{1}{3} \times 0\right) + \left(\frac{1}{3} \times \frac{1}{3} \times (-1)\right) + \left(\frac{1}{3} \times \frac{1}{3} \times 1\right) +$$

$$\left(\frac{1}{2} \times \frac{1}{3} \times 1\right) + \left(\frac{1}{2} \times \frac{1}{3} \times 0\right) + \left(\frac{1}{2} \times \frac{1}{3} \times (-1)\right) +$$

$$\left(\frac{1}{6} \times \frac{1}{3} \times (-1)\right) + \left(\frac{1}{6} \times \frac{1}{3} \times 1\right) + \left(\frac{1}{6} \times \frac{1}{3} \times 0\right) = 0.$$

And, in fact, Max would have obtained exactly this same result had he chosen *any* set of likelihoods for the three alternatives. From this fact, we conclude that Max cannot improve his average return against the equal-likelihood mixture of Min. So the strategy of mixing the choices with equal likelihood is an *equilibrium* point for the game, in the same sense that the minimax point is an equilibrium for a game having a saddle point. Thus, using a strategy that randomizes their choices, Max and Min can each announce his or her strategy to the other without the opponent being able to exploit this information to get a larger average payoff for himself or herself.

The questions that arise now are the following:

- How do we know optimal mixed strategies exist for every two-person, zero-sum game? We have seen that such a strategy exists in the Rock-Paper-Scissors Game. But perhaps there's something special about that game, and that in general there are no such optimal mixed strategies.

- If such optimal mixed strategies do indeed exist, how can we calculate them? The symmetric structure of the payoff matrix for the Rock-Paper-Scissors Game suggests strongly that the "right" thing to do for each player is to mix the alternatives R, S, and P with equal likelihood. But if the payoffs favor some actions more than others, it stands to reason that an optimal strategy will pick up on this fact, tending to play those high-payoff actions more frequently than those less favored. So how do we calculate the best way to weight the various choices when presented with a particular payoff matrix?

These were the questions that John von Neumann answered in 1928 with his celebrated Minimax Theorem, the pivotal result in the mathematical theory of games. It asserts that for every two-person, zero-sum game there is an optimal mixed strategy for each player.

## The Minimax Theorem

Let's briefly look at why the Minimax Theorem is so important. If a game has a saddle point, as in the Battle of the Bismarck Sea, then each player's choice of what to do is clear: take the action that leads to the saddle point. This ensures the best possible payoff for each player, *assuming* each acts in his or her own selfish best interest. Furthermore, each player can announce his or her saddle-point decision in advance, without the opponent's being able to exploit this knowledge to get a better payoff. But these properties are consequences of the fact that at a saddle point the maxmin equals the minmax. If the game has no saddle point, then these properties disappear, and we face a situation in which there is no clear-cut optimal way for each player to proceed.

The Minimax Theorem shows us how to restore the notion of a rational course of action in games lacking a saddle point. The Minimax Theorem says that the way to recapture rationality is to use strategies that randomize a player's choice over all possible actions. If the players proceed in this fashion, the Minimax Theorem guarantees the existence of a set of probabilities for each player, such that if each weighs his or her actions according to these probabilities he or she will each obtain the same average payoff—and this will be the best payoff each can expect to receive if the opponent plays rationally. Finally, the players can disclose their probability sets in advance without harming their own interests in any way. Let's take a longer look at this major result of 20th-century mathematics.

Following up work in 1921 by the French mathematician Emile Borel on the case of two-person, zero-sum games in which the players have no more than four actions (that is, pure strategies) at their disposal, Ernst Zermelo conjectured that there should exist mixed strategies for each player that would give them both the same expected payoff, and that this should hold independent of the number of actions each player has available. Furthermore, Zermelo claimed that this common return was the very best that the players could hope for under the risk-averse criterion that governs what game theorists mean by "rational play."

We have already seen that this must necessarily be the case for games having a saddle point, and that in this case both players have an optimal pure strategy that leads to the common payoff. In game-theoretic jargon, this common optimal payoff is usually termed the *value* of the game. But neither the geometry nor the algebra is so simple and clear-cut for games lacking a saddle point. And it was not until 1928 that John von Neumann put the entire situation on a solid footing by presenting an airtight mathematical proof validating Zermelo's conjecture. Since the *Minimax Theorem* forms the foundation upon which much of the mathematical theory of games rests, let's take a few paragraphs to describe it in more detail.

Von Neumann's Minimax Theorem makes the strong assertion that there always exists at least one mixed strategy for each player, such that the average payoff to each player is the same when they use these strategies. Moreover, this average payoff is the best return that each player can hope to obtain if his or her opponent plays rationally. Here is a slightly more formal statement of this pivotal result:

**MINIMAX THEOREM**   *For every two-person, zero-sum game there exists a mixed strategy for each player, such that the expected payoff for both is the same value* V *when the players use these strategies. Furthermore,* V *is the best payoff each can expect to receive from a play of the game; hence, these mixed strategies are the optimal strategies for the two players to employ.*   ∎

What this means is that there exists a set of probabilities for each player such that if they weight their actions in accordance with these probabilities, they will each receive exactly the same average return $V$ (actually the expected payoff is $-V$ to the minimizing player, since the game is zero-sum).

We can illustrate this result by considering again the Rock-Paper-Scissors Game given above. There we claimed (without proof!) that the optimal mixed strategy for each player was to equally weight the three alternative P, S, and R. Using these probabilities, we found that the average payoff to each player was $V = 0$. The hard part of von Neumann's result is to show that neither player can get a better expected payoff by deviating from the probabilities specified by the minimax strategy, a fact that we proved by direct calculation in the Rock-Scissors-Paper Game. Interestingly enough, to prove the Minimax Theorem, von Neumann had to develop an important extension to the fixed-point results established earlier by L. E. J. Brouwer (see Chapter 2). Readers wishing to see some of these details are urged to consult the material cited for this chapter in the bibliography.

So just as the introduction of complex numbers restored the solvability of any polynomial equation, the idea of a mixed strategy, coupled with the Minimax Theorem, restores the solvability of any two-person, zero-sum game by ensuring the existence of an equilibrium point—but now in a space of mixed rather than pure strategies. By extending the notion of what we mean by a strategy from the choice of a single course of action (a pure strategy) to a randomization over all possible actions (a mixed strategy), von Neumann succeeded in establishing the existence of a rational choice that either player can announce in advance without giving his or her opponent the slightest advantage; neither player can improve upon the return by a unilateral departure from the optimal mixed strategy. It's no wonder that von Neumann could later remark that, "As far as I can see, there could be no theory of games . . . without that theorem . . . I thought there was nothing worth publishing until the 'Minimax Theorem' was proved." Now let's look at a simple example showing mixed strategies in action.

## Fighters and Bombers

In World War II, fighter pilots normally attacked bombers by swooping down on them from the direction of the sun, a ploy known as the "Hun-in-the-Sun" strategy. But when every plane started employing this strategy, the bomber pilots simply put on their sunglasses and stared into the sun looking for the fighters. Thus, a second strategy emerged, involving an attack straight up from below. This alternative proved very effective when the fighter plane was not spotted, but was invariably fatal to the fighter pilot if he was seen, since planes go much slower when climbing than when diving. As this strategy is almost the direct opposite of the Japanese kamikaze style of attack, we label it the "Ezak-Imak" strategy ("kamikaze" spelled backward). So we now have a two-person, zero-sum game between the fighter pilots and the bomber crews. The fighters can employ either the Hun-in-the-Sun or the Ezak-Imak strategies, while the bomber crews can either look up or look down through the gunner's turret. If we agree to measure the payoff to the fighter pilot as being the chances of survival in a single mission, then we might describe the game-theoretic situation by typical survival probabilities as given in the following payoff matrix:

|  |  | Bomber Crew | |
|---|---|---|---|
|  |  | **Look Up** | **Look Down** |
| Fighter Pilot | **Hun-in-the-Sun** | 0.95 | 1 |
|  | **Ezak-Imak** | 1 | 0 |

This is clearly a game without a saddle point (why?). So there are no pure strategies that the fighter pilot and the bomber crew can choose that cannot be exploited by their opponent if the opponent gains knowledge of the choice. Instead, both parties have to mix their actions, sometimes doing one thing, sometimes the other; they must randomly select an action from among the possibilities open to them on any given sortie. Of course, this does not mean that the choices should be taken with equal likelihood. The payoffs to the fighter pilot show that the Hun-in-the-Sun strategy is almost always successful, while the Ezak-Imak strategy is quite risky, since it involves certain death if the bomber crew happens to look down. Thus, intuition suggests that the optimal mix between Hun-in-the-Sun and Ezak-Imak will strongly favor the former, with the fighter pilot only occasionally choosing the risky Ezak-Imak action just to keep the bomber crew honest. By the methods we'll discuss later, it turns out that the optimal mixed strategy for the fighter pilot is to use the Hun-in-the-Sun strategy 20 times out of every 21 sorties.

The fighter pilot might implement this strategy by placing 20 white balls and 1 black one into a bag and shaking them up. Then before each mission he would pull one ball out of the sack, a white ball signifying an attack out of the sun (the Hun-in-the-Sun strategy), the black ball telling the pilot to attack from below. It turns out that the optimal mix between looking up and looking down for the bomber crew is to look down 20 times out of 21. It's important here to note that these ratios are to be employed *on the average,* so that the strategy for the bomber crew, for instance, does *not* mean that if they have looked down the past 20 times, they should therefore look up this time. Both the fighter pilot and the bomber crew must each select a ball from the bag *before* beginning each of their flights, and then follow the action dictated by the color of the ball they have drawn.

By using the methods of the next section, these mixed strategies for the fighter pilot and the bomber crew lead to an expected safety level of 0.9524, slightly higher than the certain safety level of 0.95 that the fighter pilot has by always following the Hun-in-the-Sun strategy. Of course, to obtain this slightly higher expected level of survival, the fighter pilot has to accept a chance of certain death should both he and the bomber crew happen to pull the black ball (with probability $1/21 \times 1/21 \approx 0.0023$). Let's now look at where these numbers come from.

## Computing Optimal Mixed Strategies

We consider here only games where one player or the other (but not necessarily both) has at most two actions available (that is, two pure strategies). The procedure for calculating optimal mixed strategies in the general case in which both players have more than two pure strategies is a bit more complicated than we can describe here. The reader can consult the literature cited in the bibliography for details of the general situation.

To see what's involved in the calculation of optimal mixed strategies for these simpler games, let's look at the following payoff matrix for a game in which Player A has two pure strategies, while Player B has three:

|  |  | Player B |  |  |
|---|---|---|---|---|
|  |  | **B-I** | **B-II** | **B-III** |
| Player A | **A-I** | 0 | 5/6 | 1/2 |
|  | **A-II** | 1 | 1/2 | 3/4 |

Suppose Player A's optimal mixed strategy is to play strategy A-I $x$ percent of the time and play A-II a fraction $1 - x$. The expected, or average, payoffs to Player A against the three strategies of Player B are then

$$0x + 1(1 - x) = 1 - x \qquad \text{against strategy B-I,}$$

$$\frac{5}{6}x + \frac{1}{2}(1 - x) = \frac{1}{2} + \frac{1}{3}x \qquad \text{against strategy B-II,} \qquad (*)$$

$$\frac{1}{2}x + \frac{3}{4}(1 - x) = \frac{3}{4} - \frac{1}{4}x \qquad \text{against strategy B-III.}$$

It's instructive to show these results graphically by plotting each expected payoff as a function of the quantity $x$, the fraction of the time Player A uses strategy A-I. Each of the curves representing these expected payoffs is just a straight line, as shown in Figure 1.3. For each value of $x$, the height of the lines at that point represents the payoff to Player A against each of Player B's three strategies when Player A chooses A-I a fraction $x$ and A-II a fraction $1 - x$ of the time.

Player A is concerned about the smallest payoff he or she could receive for each possible value of $x$ that might be chosen. Geometrically, this smallest expected payoff is the point on the lowest of the three lines at each value of $x$. So Player A will try to choose $x$ so as to make this minimum payoff as large as possible. In other words, A will look for the value of $x$ that corresponds to the highest point on the lowest of the three lines. This occurs at the maxmin point, which is at $x = Q$ in the figure. This corresponds to the highest point $P$ on the lower envelope of the

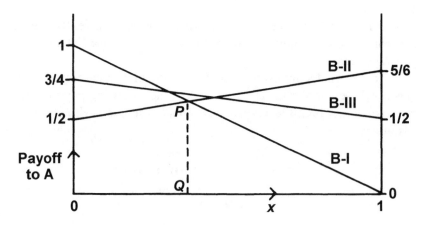

**Figure 1.3**   Graphical solution of the game.

three lines. The distance $PQ$ represents the value of the game, which we earlier denoted by $V$, while the distance $x = 0Q$ is the fraction of the time Player A should employ strategy A-I so as to obtain this optimal expected payoff. And, of course, the distance $1 - x = 1 - 0Q$ is then the fraction of the time that Player A should use strategy A-II.

It's important to note here that the line representing strategy B-III lies entirely above the lower envelope of the three strategy lines. This means that the pure strategy B-III is not worth considering for Player B, since playing this strategy against the optimal mixed strategy of Player A leads to a payoff for Player B that is no better than what could be obtained by playing some combination of the other two strategies. This is bad for Player B, who is the minimizing player and thus wants to get the smallest possible payoff, not the largest. Therefore, Player B can drop strategy B-III from consideration (this shows up in the fact that this strategy appears with probability zero in Player B's optimal mixed strategy).

From the geometry of the situation, it's fairly evident that the maxmin point typically lies at the intersection of only two of Player B's strategy lines, since it would take a very special set of circumstances for three or more lines to happen to meet at a single point. But suppose line B-III in Figure 1.3 were moved down so that it passed through the point $P$. In this case strategy B-III would still not be part of Player B's optimal mixed strategy, since using that strategy confers no advantage to Player B over using a combination of pure strategies B-I and B-II. Of course, if line B-III is lowered still further, the maxmin point changes to the intersection of B-II and B-III, in which case strategy B-I now drops out of consideration. This situation is completely general; the maxmin point for Player A will always lie at the intersection of two pure strategies of Player B. This enables us to reduce the calculation of the optimal mixed strategies for both players to that of a $2 \times 2$ game.

Usually we want a more precise solution to the game than can be read off from the graphical answer shown in Figure 1.3. The way to compute the exact mixed strategies for the two players is to make use of the game-theoretic fact noted earlier that when Player A uses the optimal mixed strategy against either of the pure strategies B-I or B-II, Player A must receive exactly the same expected return, namely, what we've called the value of the game. There was an example of this in the Rock-Paper-Scissors Game, in which Max received the same average return using *any* mixed strategy—just as long as Min continued to play her optimal mixed strategy. From the equations $(*)$ given above, we see that the two expected payoffs for Player A against the pure strategies B-I and B-II are

$$E(\text{against B-I}) = 1 - x,$$
$$E(\text{against B-II}) = \frac{1}{2} + \frac{1}{3}x.$$

Equating these two expressions yields $x = 3/8$. Substituting this value into either of the above relations leads to the value of the game as $V = 5/8$. Thus, the optimal mixed strategy for Player A is to play strategy A-I $3/8$ of the time, while playing A-II the remaining $5/8$ of the time. By following this strategy, Player A will receive an expected payoff of $5/8$. By a similar line of argument, it can be shown that Player B has the optimal mixed strategy consisting of playing B-I $1/4$ of the time, B-II $3/4$ of the time, and never playing strategy B-III. An easy calculation shows that with this mix, Player B will receive an expected payoff of $-5/8$—exactly what we expect for a zero-sum game.

## Game Theory—A Taxonomy

According to Caesar, all Gaul was divided into three parts. We can make a similar division in the mathematical theory of games. The first, most classical, and by far, best-developed sector is the class of games we've been discussing thus far, the zero-sum games. As we've seen, there is a clear-cut notion of what constitutes a solution to such games, the mixed strategies from the Minimax Theorem. This solution is not only unique but has the inestimable attraction that it provides a mathematical expression of the idea of "rational action." This notion agrees with at least one interpretation of our everyday understanding of the term "rational," namely, a view of rationality that we might colloquially describe as "hope for the best, prepare for the worst."

From a mathematical point of view, the Minimax Theorem essentially closed out the study of zero-sum games, providing something akin to a complete theory of action for such encounters. Of course, the practical problems of actually implementing an optimal mixed strategy—or even verifying the assumptions of the situation—may be considerable. But there is no mystery about what's to be done—not so for the remaining branches of game theory. For the sake of completeness, let's briefly consider these other two branches in turn.

In 1994, for the first time in the 93-year history of the Nobel prizes, an award was given for work done solely in pure mathematics: The Nobel Memorial Prize in Economic Science was shared by Princeton mathematician John Nash with John Harsanyi and Reinhard Selten for seminal work in game theory. Basically, what Nash did was prove a theorem that generalizes the Minimax Theorem to the case of nonzero-sum games involving two or more players when the players are in direct competition—the so-called *noncooperative games*. To understand Nash's achievement, let's reexamine the notion of a solution to a game.

Consider the situation facing the maximizing player. Suppose the player examines each available pure strategy, singling out the minimum payoff that could be obtained using that strategy against each of the pure strategies of his or her opponent. The maximizing player then chooses the strategy that yields the maximum of these minimum possible payoffs. This is what we call the *maxmin* strategy. A completely analogous argument leads to the *minmax* strategy for the minimizing player. These two strategies taken together are called a *minimax pair* of strategies for the game.

Now let's look again at the idea of an *equilibrium pair* of strategies. This is simply a pair of strategies such that a player who deviates unilaterally from his or her equilibrium strategy in this pair will receive a worse expected payoff than by not deviating. An important corollary of the Minimax Theorem is that these two rather different concepts of a strategy coincide in the case of zero-sum games: an equilibrium pair is a minimax pair, and conversely, but not so when the game is nonzero-sum. Nash's Theorem addresses this situation. Here is an informal statement of his result.

---

**NASH'S THEOREM**   *Any* n-*person, noncooperative game (zero-sum or nonzero-sum) for which each player has a finite number of pure strategies has at least one equilibrium set of strategies.*   ■

We note three aspects of this result: (1) it generalizes the Minimax Theorem in the sense that it establishes the existence of an equilibrium solution for both zero-sum and nonzero-sum games, (2) it shows that there may be more than one such solution, and (3) it extends to the case of any finite number of players. The second point, incidentally, leads to a great deal of difficulty in trying to match up the various possible Nash equilibria with anything that looks even remotely like what we would consider "rational" behavior in daily life. I'll leave it to the reader to consult the material cited in the bibliography for a more detailed discussion of this and other aspects of equilibrium strategies.

The foregoing result by Nash pertains to noncooperative games. But as we're going to see shortly, there may be gains for the players if they agree to cooperate—at least partially—rather than go for each other's throats. This leads to the idea of a *cooperative game,* in which the players can form various coalitions in order to improve their respective payoffs, or, perhaps, make side payments to other players in order to influence their actions.

The situation in terms of finding a completely satisfactory solution concept for such *n*-person, cooperative games is a murky one. Beginning with the pioneering work of John von Neumann and Oskar Morgenstern and continuing on down through efforts by Martin Shubik, Lloyd Shapley and others, a number of solution concepts have been advanced for such games—equilibrium sets, the core of a game, and the Shapley value, to name but a few. Some of these (like the Shapley value) are unique, while others like the core of a game are not. And so far no one of these approaches to a solution have gained a consensus of the game theory community as constituting a completely satisfactory way of characterizing "rational" action on the part of players in games when at least some degree of cooperation is allowed. But to explore these matters would take us too far afield in an introductory volume of this sort, so I'll leave it to the reader to investigate these various issues in the material cited in the bibliography. For now, let's return to a consideration of two-person games when the total payoffs to the players does not sum to zero.

The Minimax Theorem applies to games in which there are just two players and for which the total payoff to both parties is zero, regardless of what actions the players choose. The advantage of these two properties is that with two players whose interests are directly opposed we have a game of pure competition, which allows us to define a clear-cut mathematical notion of rational behavior that leads to a single, unambiguous rule as to how each player should behave.

Allowing more than two players into the game and/or postulating payoff structures in which one player's gain does not necessarily equal the other player's loss brings us much closer to the type of games played in real life. Unfortunately, it's generally the case that the closer you get to the messiness of the real world, the farther you move from the stylized and structured world of mathematics. Game theory is no exception. By admitting either of these new features into the game, we open up the possibility of at least partial *cooperation* among the players. From a mathematical point of view this creates enormous difficulties, since in such situations it's no longer possible, in general, to offer a clear-cut mathematical argument for how a rational player should behave. Actually, it's no longer possible to even agree on what we *mean* by "rational" behavior in such cases. But this is not to suggest that nothing can be done. And, in fact, a considerable amount of effort has been expended over the past 50 years or so in trying to produce a coherent theoretical framework within which to speak of such games, a bit of which was outlined above. So in the remaining sections of this chapter we shall explore a few of these extensions by looking at some specific games illustrating the fascinating—and tricky—aspects of cooperative versus competitive behavior.

We start by considering the effects of changing the game's payoff structure to allow for returns for which one player's gain is not equal to the other player's loss. This simple change opens up the possibility that the players may sometimes be better off by cooperating than by competing.

# Chicken

A well-known game, whose underlying principles date back at least as far as the Homeric era, involves two motorists driving toward each other on a collision course. Each has the option of being a "chicken" by swerving to avoid the collision (C), or of continuing on the deadly course (D). If both drivers are chicken, they each survive and receive a payoff of, say, 3 units. But if one "chickens out" and the other drives on, the chicken loses face (but not his or her life) and the "macho" driver wins a prestige victory. In this case, the chicken receives 2 units, whereas the other driver receives 4 points. Finally, if they both carry on to a fatal collision, they each receive Death's reward of 1 unit. Readers who came of age in the 1950s will recognize this as the game played by James Dean against his high-school nemesis Buzz in Nicholas Ray's 1955 cult film *Rebel Without a Cause*. The only difference is that in the film the cars were heading toward a cliff, the winning driver being the one to dive out of his car last before it went over the cliff.

The payoffs for the Chicken game are listed in the matrix below, in which to honor the Hollywood version of the game we have named the players Jimbo and Buzz. The first entry in each pair of payoffs goes to Jimbo, while the second entry is the number of points received by Buzz.

|  |  | Buzz | |
|---|---|---|---|
|  |  | **C** | **D** |
| Jimbo | **C** | (3, 3) | (2, 4) |
|  | **D** | (4, 2) | (1, 1) |

The first thing we note is that the sum of the payoffs is not the same constant for all possible choices. So it *might* be beneficial for the players to consider forsaking a purely competitive, "go for the jugular" attitude and cooperate—at least partially—instead. Moreover, we see immediately that there is no dominant strategy in this game for either player. What this means is that there is no single course of action for either player that will yield the largest payoff regardless of the decision taken by the other player. So, for instance, we see that the swerving strategy C for Jimbo is best if Buzz plays the deadly strategy D of staying on a collision course. But it is not as good as D if Buzz himself decides to swerve by choosing C. So the decision to swerve off the road is not dominant for Jimbo. Similar arguments apply to the pure strategies of

27

either always swerving or always continuing on the collision course for the two drivers. So there is no dominant strategy for either player in the game of Chicken.

To analyze what the drivers should do, let's first look at the game from Jimbo's standpoint. He can argue as follows: "My minimum payoff if I swerve is 2 units, while if I continue to drive straight on my minimum return is 1. Thus, taking the maximum of these two possibilities, my rational action is to choose C and swerve off." A similar analysis by Buzz leads to the same conclusion. Hence, the minimax strategies for the two results in both drivers "chickening out"—exactly what would have happened in the film if Buzz hadn't caught the sleeve of his jacket in the door handle and thus been prevented from jumping out of his car before it went over the cliff as he was trying to do. Note here how an exploiter who deviates from this chickening-out strategy can gain an advantage, and by doing so invariably affects the other driver adversely by such a deviation. So not only does the deviator harm the other player, it also puts itself and the other player in a position where they may have a disastrous outcome.

Now look at the game from the standpoint of equilibrium strategies. It's clear by inspection of the payoffs that if Buzz and Jimbo choose either of the pair of actions (C, D) or (D, C), neither player will have an incentive to depart unilaterally from the choice. Thus, the equilibrium strategies differ from the minimax ones—unlike the situation in two-person, zero-sum games.

Chicken has the peculiar feature that it is impossible to avoid playing the game with someone who insists, since to refuse to play is effectively the same as playing and losing. In addition, the player who succeeds in making a commitment to the dangerous D option appear convincing will *always* win at the expense of the other player, assuming the other player is rational. So a player who has a deserved reputation for hot-headed recklessness enjoys a decided advantage in Chicken over one who is merely rational. Perhaps this accounts for the aversion of many academics to this kind of irrational winning strategy, as most professorial types seem to pride themselves on being both risk-averse and ultra-rational men and women, an unhappy losing combination for those engaged in the game of Chicken. This becomes especially true if Chicken is played a number of times, because the player who gains an early advantage usually maintains (or even increases) that advantage later on. Once such a player has successfully exploited the opponent, this player gains confidence in his or her ability to get away with the risky strategy in the future while making his or her opponent all the more fearful of deviating from the cautious minimax alternative.

To show that the payoff structure and decision analysis underlying Chicken is no academic curiosity, we now look at a real-life situation in which world leaders faced the same kind of choices and payoff structure displayed by Chicken.

## The Cuban Missile Crisis

Without a doubt, the most dangerous confrontation between world powers of this century took place in October 1962 when the USSR attempted to place nuclear-armed missiles in Cuba. When the presence of these missiles was confirmed on October 14, President John F. Kennedy convened a so-called Executive Committee of high-level officials to decide on a course of action for the United States. After consideration of several alternatives, the ExecComm narrowed the options to two: a naval blockade or an air strike. By the same token, Soviet Premier Nikita Khruschev also had two choices open to him: withdraw the missiles or leave them in place. In game-theoretic terms, we can summarize the situation facing Khruschev and Kennedy with the following payoff matrix:

|  |  | USSR | |
|---|---|---|---|
|  |  | **Withdrawal** | **Maintain** |
|  | **Blockade** | (3, 3) | (2, 4) |
|  |  | *(Compromise)* | *(USSR Victory)* |
| United States |  |  |  |
|  | **Air Strike** | (4, 2) | (1, 1) |
|  |  | *(U.S. Victory)* | *(Nuclear War)* |

Here the first entry in each payoff is the preference ranking of that action by the United States, while the second is the Soviet Union's ranking, where 4 is best, 3 next best, and so on. With these rankings, the Cuban Missile Crisis has the same payoff structure as the game of Chicken. And, indeed, the basic conception that most people have of this crisis is that the two superpowers were on a "collision course." As everyone knows by now, the decisions actually taken were blockade and withdrawal, resulting in a compromise that resolved the so-called Cuban Missile Crisis. So, although in one sense the United States "won" the game by getting the Soviets to remove their missiles from Cuba, the Soviets also managed to extract a promise from President Kennedy not to invade Cuba, suggesting that the final resolution of the Crisis was actually a compromise of sorts.

It goes without saying, I hope, that the strategy choices and preferences given in the payoff matrix offer only a skeletal outline of the Cuban Missile Crisis itself, which unfolded over the now-famous "thirteen days in October." First of all, both sides considered more than the two alternatives listed in the matrix, as well as several variations of each alternative. For instance, the Soviets demanded withdrawal of U.S. missiles from Turkey, a demand ignored by the American side. Furthermore, the likely outcomes given in the payoff matrix may not have been very "probable" at all. To illustrate this crucial point, consider the combination of a U.S. air strike and a Soviet withdrawal. While we have labeled this a U.S. victory, it may well have been the case that the Soviet Union would have regarded an air strike on their missiles as threatening their national interests. So this combination of actions could easily have ended in a nuclear war, giving it the same value as the Air Strike and Maintain decision pair. Furthermore, even though the Soviets responded to the blockade by removing their missiles, the United States held out the possibility of escalating the conflict to at least an air strike if the blockade did not achieve its objective. This suggests that the decision to blockade was not to be thought of as in any way final, and that the United States considered its strategy choices still open even after imposing the blockade.

## Mixed-Motive Games

Chicken is the most well-known of an entire class of games in which cooperative behavior may benefit both players—under the right circumstances. Since it's a tricky business determining when to cooperate and when to compete, we shall spend a few pages describing a few more games of this sort, in order to get a feel for the kinds of situations that can arise in which cooperation is favored over competition.

Examination of the Chicken Game in both its Hollywood and Cuban Missile Crisis forms shows that there is no single course of action

30

that is best for each player, independent of what the other player decides to do. This means, for example, that the interests of both Kennedy and Khruschev were neither diametrically opposed nor directly parallel. Such games are often termed *mixed-motive* games, reflecting the possibility that both players may want the same outcome, that is, they may want to cooperate rather than compete.

Furthermore, the Cuban Missile Game has no equilibrium point representing a dominant course of action for each party. In contrast, the game with the following payoff matrix

|  |  | Player II | |
|---|---|---|---|
|  |  | C | D |
| Player I | C | (4, 4) | (2, 3) |
|  | D | (3, 2) | (1, 1) |

has a dominant strategy C for each player. Such games are, of course, uninteresting from a strategic point of view since it's clear by inspection what a rational player should do.

It turns out that there are only four types of nontrivial mixed-motive games, of which Chicken is the prototypical representative of one such type. For the sake of completeness and subsequent discussion, we now briefly describe the three remaining types of "interesting" mixed-motive games. For consistency and ease of exposition, we shall suppose that in these games each player has two possible actions, which, for reasons that will become apparent later, we label "C" and "D."

## Leader

Consider the case of two drivers attempting to enter a busy stream of traffic from opposite sides of an intersection. When the cross traffic clears, each driver must decide whether to concede the right of way to the other (C), or drive into the gap (D). If both concede, they will each be delayed, whereas if they drive out together there may be a collision. However, if one drives out while the other waits, the "leader" will be able to carry on with his or her trip, while the "follower" may still be able to squeeze into the gap left behind by the leader before it closes again. A typical payoff matrix for this Leader Game is:

Driver II

|  | | C | D |
|---|---|---|---|
| Driver I | C | (2, 2) | (3, 4) |
|  | D | (4, 3) | (1, 1) |

Again we see that there is no dominant strategy. According to our by-now-familiar minimax arguments, to avoid the worst possible outcome each driver should choose action C, thereby ensuring that neither will receive a payoff less than 2 units. However, the minimax strategies for the two players are not in equilibrium, since each driver would have reason to regret his or her choice when he or she sees what the other driver has done. This simple observation shows that the minimax principle cannot be used as a basis for prescribing rational courses of action in mixed-motive games.

In fact, it turns out that there are *two* equilibrium strategies in the Leader Game: the pair (C, D), in which Driver I concedes the right of way so that Driver II drives into the gap, and its opposite, the pair (D, C). These actions appear at the off-diagonal corners of the payoff matrix. If Driver I chooses action D, the second driver can do no better than to choose C, and vice versa. In other words, neither can do better by deviating from an equilibrium outcome. However, in contrast to zero-sum games, in which such equilibrium points are always equivalent, in the Leader Game Driver I prefers the (D, C) equilibrium, while Driver II prefers (C, D). There is no mathematical way of settling this difference, which is why in real-world situations of this sort the impasse is often resolved by the fact that one of the equilibrium points is more visible to the drivers than the other. For example, cultural and/or psychological factors may enter to break the deadlock, with various tacit rules like "first come, first served" or signaling schemes like the blinking of lights being used to single out one of the equilibria. This kind of signaling, incidentally, is in sharp contrast to the situation in zero-sum games, where such signals would definitely not be to a driver's advantage, since it would immediately give useful information to the other driver.

## Battle of the Sexes

In this game, a married couple has to choose between two options for their evening entertainment. The husband prefers one type of entertainment, say a movie, whereas his wife prefers another, for instance, going out for a pizza. The problem is that they would both rather go out together than alone. If they each opt for their first choice (call it action C), they end up going out alone and each receives a payoff of 2 units. But if they each make a sacrifice and go to the activity they don't like (action D), each suffers, and they receive a payoff of only 1 unit apiece. But if one sacrifices while the other gets their first choice, then they still go out together, but the "hero" who sacrifices receives 3 units of "reward," while the other party gets 4. The payoff matrix for this game then turns out to be:

|  |  | Wife | |
|---|---|---|---|
|  |  | C | D |
| Husband | C | (2, 2) | (4, 3) |
|  | D | (3, 4) | (1, 1) |

Battle of the Sexes has a number of features in common with Leader: (1) neither the husband nor the wife has a dominant strategy, (2) the minimax strategies intersect in the nonequilibrium result (C, C), and (3) both strategies (C, D) and (D, C) are in equilibrium. But, in contrast to the Leader Game, in the Battle of the Sexes Game the player who deviates unilaterally from the minimax strategy rewards the other player more than himself or herself. This is just the opposite of what happens in the Leader Game, where the deviator receives a greater reward than the opponent. But just as in the Leader Game, here also a player can gain by communicating with the other player in order to obtain some level of commitment to action C. So, for instance, the husband might announce that he is irrevocably committed to his first choice of going to the movies, in which case this will work to his advantage if his wife then acts so as to maximize her own return. The only difficulty lies in convincing her that he's serious. The main point to note here is that some kind of commitment is needed in order for both parties to achieve the best possible joint outcome in the Battle of the Sexes Game.

## The Prisoner's Dilemma

The last basic type of mixed-motive game, and by far the most interesting, is the well-chronicled game involving two prisoners who are each accused of a crime. Both prisoners have the option of concealing information from the police (C) or disclosing it (D). If they each conceal the information (that is, they cooperate), they will both be acquitted with a payoff of 3 units to each. If one conceals and the other "squeals" to the police, the squealer receives a reward of 4 units, while the payoff to the "martyr" is only 1 unit, reflecting his or her role in the obstruction of justice. Finally, if they both talk, they will each be convicted—but of a lesser offense—thereby receiving a payoff of only 2 units apiece. An appropriate payoff matrix for the Prisoner's Dilemma Game is

|  |  | Prisoner II | |
|---|---|---|---|
|  |  | C | D |
| Prisoner I | C | (3, 3) | (1, 4) |
|  | D | (4, 1) | (2, 2) |

(Note: Our use of the symbols C and D to represent the possible actions by the players in all of these mixed-motive games is motivated by the usual interpretation of the actions in the Prisoner's Dilemma Game. Here C represents "cooperating" with your pal and not confessing, whereas D signifies "defecting" to the police and giving the information needed for a conviction. This interpretation is due to Princeton game theorist Albert Tucker, who coined the term "Prisoner's Dilemma" in the late 1950s, following introduction of the dilemma by Merrill Flood of The RAND Corporation in 1951.)

The Prisoner's Dilemma Game is a real paradox. The minimax strategies intersect in the choice of mutual defection, which is also the only equilibrium point in the game. So neither prisoner has any reason to regret a minimax choice if the other also plays minimax. The minimax options are also dominant for both prisoners, since each receives a larger payoff by defecting than by cooperating when playing against either of the other player's choices. Thus, it appears to be in the best interests of each prisoner to defect to the police—*regardless* of what the other player decides to do. But if both prisoners choose this individually rational course of action, the 2 units they each receive are less than the 3 units they could each have pocketed if they had chosen to remain silent.

The essence of the paradox in the Prisoner's Dilemma lies in the conflict between individual and collective rationality. Following individual rationality, it's better for a prisoner to defect and give information to the police. But, paradoxically, if both attempt to be "martyrs" and remain silent, they each wind up being better off. What's needed to ensure this better outcome for both players is some kind of selection principle based on their joint interests. Perhaps the oldest and most familiar such principle is the Golden Rule of Confucius: "Do unto others as you would have them do unto you." Note, however, that the Golden Rule can be disastrous in other kinds of games. For example, if both the husband and the wife employ the Golden Rule in Battle of the Sexes, the outcome is the worst possible and leads to each of them going out alone to an entertainment they don't want.

Our deliberations thus far have centered solely on what one might term "one-off" games; that is, games in which the players make a single decision once and for all, take the corresponding action, and more or less let the chips fall where they may. In short, there's only a single play of the game. And certainly many games in life are of exactly this type. But many more are not. So now it's time for us to look at the ways in which things change when we are faced with conflict situations in which there may be a sequence of plays, perhaps even a game that goes on indefinitely. And just to see that things do indeed change, consider again the Prisoner's Dilemma Game.

It's a fairly easy exercise to verify that defection by both players is the equilibrium strategy to employ for not only a single play of the game, but also for any *finite* sequence of plays in which the number of plays is known to both players beforehand. To see this, consider a game of $N$ plays. By our earlier arguments, both players should defect if the game is played only once. So in a sequence of $N$ plays, they should both defect on the last play since there are no plays remaining. But if they both defect on the $N$th play, they are faced with a game involving a sequence of $N - 1$ plays. Employing the same line of reasoning, both defect on the last play of this sequence, play $N - 1$. And so it goes, clear back to the very first play, one pair of defections followed by another and another and another. Thus, the more profitable alternative C never has a chance to get started in such a situation.

By the arguments we'll explore in a moment, as soon as we admit the possibility that the game may continue indefinitely, cooperation becomes a possibility. What "indefinitely" means here, of course, is that after any play of the game there is a chance that the players may interact again, or what is the same thing, that $N$ is not known to the players *in advance*. But we're getting a little ahead of ourselves, as these matters are more properly the subject of the next few sections. So without further ado, let's leave the realm of the one-off game and talk about *multistage* games instead.

## The Emergence of Cooperation

The cornerstone of sociobiological reasoning is the claim that human behavior patterns, including what look on the surface to be selfless acts of altruism, emerge out of genetically selfish acts. In the context of the Prisoner's Dilemma Game, we can translate this sociobiological thesis into the statement that the individually rational act of defection will always be preferred to the collectively rational choice of cooperation. The question then becomes whether there are any circumstances that would enable a band of cooperators to get a foothold in a population of egoists. If there is no way for cooperative acts to emerge naturally out of a kind of enlightened self-interest, it's going to be very difficult for the sociobiologists to support the case for "selfish gene" theories.

Couched in the terms of evolutionary game theory, the strategy of always defecting is what's called an *evolutionary stable strategy* (ESS), since players who deviate from this strategy can never make inroads against a population of defectors. Or can they? Are there any situations in which a less cutthroat course of action can ultimately establish itself in a population of defectors? This was the question that Robert Axelrod set out to answer in one of the most intriguing psychological experiments carried out in recent years. The separate issues that Axelrod wanted to address were: (a) How can cooperation get started at all in a world of egoists? (b) Can individuals employing cooperative strategies survive better than their uncooperative rivals? (c) Which cooperative strategies will do best and how will they come to dominate?

Axelrod began his experiment by using the well-known fact in game theory that there are many equilibria for a Prisoner's Dilemma Game of indefinite duration. For example, while ALL D, the strategy of always defecting, is uninvadable for a sequence of Prisoner's Dilemma interactions of known, fixed, and finite duration, there may be alternative ESS strategies if the number of interactions is not known by both parties in advance. Put another way, if after having played a round of the Prisoner's Dilemma Game there is a nonzero chance that the game will continue for another round, then maybe there is a nice strategy that is also an ESS. Here by "nice" we mean a strategy that would not be the first to defect.

To test this idea, Axelrod invited a number of psychologists, mathematicians, political scientists, and computer experts to participate in a contest pitting different strategies against one another in a computer tournament. The idea was for each participant to supply what he or she considered to be the best strategy for playing a sequence of Prisoner's Dilemma interactions, the different strategies then competing against each other in a round-robin tournament. Fourteen competitors sent in strategies in the form of computer programs. The ground rules allowed the programs to make use of any information about the past plays of the game. Furthermore, the programs did not have to be deterministic, but were allowed to arrive at their decision by some kind of randomizing device if the player so desired. The only requirement was that the program ultimately come to a definite decision for each round of play: Cooperate (C) or Defect (D). In addition to the submitted strategies, Axelrod also included the strategy RANDOM, which took the decision to cooperate or defect by, in effect, flipping a coin. In the tournament itself, every program was made to engage every other (including a clone of itself) 200 times, the entire experiment being carried out 5 times in order to smooth out statistical fluctuations in the random-number generator used for the nondeterministic strategies.

The winning strategy turned out to be the simplest. This was the three-line program offered by Anatol Rapoport describing the strategy TIT FOR TAT. It consists of two rules: (1) cooperate on the first encounter, and (2) in subsequent rounds, do whatever your opponent did on the previous round. That such a simple, straightforward strategy could prevail against so many seemingly far more complex and sophisticated rules for playing the Prisoner's Dilemma seems nothing short of miraculous. The central lesson of this tournament was that in order for a strategy to succeed, it should be both nice and forgiving, that is, it should be willing both to initiate and to reciprocate cooperation. Following a detailed analysis of the tournament, Axelrod decided to hold a second tournament to see if the lessons learned the first time around could be put into practice to develop even more effective cooperative strategies than TIT FOR TAT.

As prelude to the second tournament, Axelrod packaged up all the information and results from the first tournament and sent it to the various participants, asking them to submit revised strategies. He also opened up the tournament to outsiders by taking out ads in computer magazines, hoping to attract some programming fanatics who might take some time off from their daily "hacking" chores to devise truly ingenious strategies. Altogether, Axelrod received 62 entries from around the world, including one from the renowned evolutionary biologist John Maynard Smith, who discovered the concept of an evolutionary stable strategy while looking for applications of game theory to biology during a sabbatical leave one year at the University of Chicago. The winner? Again it was Rapoport with TIT FOR TAT ! So even against this supposedly much stronger field, Rapoport's game-theoretic version of the Golden Rule was the hands-down winner. The general lesson emerging from the second tournament was that not only is it important to be nice and forgiving, but it's also important to be both provocable and recognizable, that is, you should get mad at defectors and retaliate quickly but without being vindictive, and your actions should be straightforward and transparent, avoiding the impression of being too complex.

Given that a strategy has become established, the question arises of how well it can resist invasion by someone playing a different strategy. It's a rather simple matter to show that once TIT FOR TAT is established in a community of players, it's impossible for any mutant strategy to dis-place it—provided the likelihood of future interactions is great enough. So if there is a good possibility that two players will face each other again, and if the dominant strategy in the community is TIT FOR TAT, then any deviation from this strategy will result in a loss of points to the mutant. In fact, we can give a precise value for how large this possibility of future interaction must be in order for TIT FOR TAT to be an ESS.

To do this, we consider the general Prisoner's Dilemma payoff ma-trix, now calling the players simply Player A and Player B. In symbolic terms, this payoff matrix is given by

|  | Player B | |
|---|---|---|
|  | **C** | **D** |
| Player A **C** | (R, R) | (S, T) |
| **D** | (T, S) | (P, P) |

Here the payoffs $P$, $R$, $S$, and $T$ are just real numbers satisfying the conditions $T > R > P > S$ and $R > (T + S)/2$. The first condition ensures that the payoffs will lead to individual self-interest (ALL D) being an ESS strategy, while the second guarantees that if the players somehow get locked into an "out of phase" alternation of cooperations and defections, each will do worse than if they had cooperated with each other on each play from the very beginning. Incidentally, the labels for the payoffs are chosen to reflect the story motivating the Prisoner's Dilemma appellation for the game that we introduced earlier in the chapter. Here $P$ is the "punishment" for joint defections, $R$ is the "reward" for jointly cooperative behavior, $S$ is the "sucker's" payoff received by a cooperator whose opponent defects, and $T$ is the "temptation" payoff for defection if the opponent cooperates.

With this payoff structure, which in essence *defines* the Prisoner's Dilemma Game, TIT FOR TAT is an ESS strategy in the iterated game if and only if the probability $w$ of two players meeting in the future is strictly greater than the larger of the numbers

$$(T - R)/(T - P) \quad \text{and} \quad (T - R)/(R - S).$$

Let's look at what these two numbers are trying to tell us. The first quantity is simply the relative payoff for being nasty and getting away with it, as opposed to being nasty and getting caught. The second ratio is the incremental difference in what you receive for being nasty and getting away with it compared with the incremental amount you receive for being nice without being suckered in. Unfortunately, words alone don't suffice to describe exactly why the larger of these numbers represent the cutoff point at which TIT FOR TAT ceases to be an ESS. We need to dive into the mathematical details for that. But on the basis of intuition alone, it's fairly evident that if either of these numbers is large, then there is a relatively large reward—hence, temptation—for being nasty. So in that case playing TIT FOR TAT is at least as good as ALL D, suggesting that TIT FOR TAT cannot be displaced by a population of defectors.

It turns out that TIT FOR TAT is not the only ESS for the iterated Prisoner's Dilemma Game. For example, ALL D is another. So how could TIT FOR TAT ever get started in a population that initially consists of all defectors? There appear to be at least two different mechanisms providing a plausible path whereby TIT FOR TAT might "infiltrate" such a fundamentally hostile environment. The first is *kin selection,* that is, helping one's relatives. Not defecting in a Prisoner's Dilemma Game is altruism of a kind, since the altruistic individual is foregoing returns that might have been taken. Thus, one way cooperation can evolve is if two players are sufficiently closely related. In effect, recalculation of the payoff matrix in such a way that an individual has an interest in a partner's success (that is, computing the payoffs in terms of inclusive fitness) can often reverse one or both of the inequalities $T > R$ and $P > S$. In that case, cooperation becomes unconditionally favored. This implies that the benefits of cooperation can be obtained by groups of sufficiently closely related players. And once the genes for cooperation exist, selection will promote strategies relying on cooperative behavior.

*Clustering* is the second mechanism by which cooperation can emerge in an essentially ALL-D environment. Suppose that a small group of individuals is using a strategy like TIT FOR TAT, and that a certain proportion $p$ of the interactions of members of this cluster is with others from the cluster. Then if playing against members of the cluster constitutes a negligible fraction of the interactions for the rest of the population, and if the probability $p$ and the likelihood of future interactions are large enough, a cluster of TIT FOR TAT individuals can become viable—even in an environment in which the overwhelming majority of the players are being nasty by defecting at every encounter. We note in passing that clustering is often associated with kinship, and the two mechanisms can often reinforce each other. But it's still possible for clustering to be effective even in the absence of kinship.

## Real Worlds, Artificial Games

In 1954, John Williams, head of the Mathematics Division of The RAND Corporation, wrote that game theorists

> are often viewed by the professional students of man as precocious children who, not appreciating the true complexity of man and his works, wander in wide-eyed innocence, expecting that their toy weapons will slay live dragons just as well as they did inanimate ones.

And this from the man who was responsible for not only writing one of the most readable introductions to the theory of games for the general reader, but who more importantly provided the initial impetus and intellectual (and financial) support for most of the mathematical work underpinning the entire edifice of the theory of games of strategy as we know it today. So it makes one pause to wonder whether or not game theory is just an idle intellectual exercise devised for keeping itinerant mathematicians employed. Or is there something worthwhile here for the rest of humankind?

I think most game theorists would be both flattered and surprised to hear that anyone takes their ideas seriously enough to think that the *theory* of games can actually provide a workable solution to an *actual* real-world conflict. The theory is just not sufficiently well developed and rests on far too idealistic assumptions to come even close to bridging the gap between the universes of mathematics and reality. This is the bad news. The good news is that game-theoretic ideas and concepts have already provided a wealth of deep and important insights into the workings of the conflict resolution process in real-world situations, some of which we have examined in this chapter.

The sine qua non of game theory is the assumption that the players act rationally, making their decisions in an essentially amoral, self-serving egotistical fashion. Philosophers, moralists, and others of like mind have taken the rationality assumption to imply that game theory portrays a world of intelligent and calculating people pursuing what each perceives to be his or her own self interest. Of course, these misgivings equate the "players" of the theory with fallible, cranky, irrational human beings. So in this sense the abstract players of the theory bear no more resemblance to real-world decisionmakers than the point particles of Newtonian mechanics bear to oblate spheroids like the Earth or Jupiter. But it's hard to argue that these idealistic assumptions about particle systems in physics are irrelevant to real-world concerns. And so it is too with game theory.

On this encouraging note, we conclude our brief excursion into the world of strategy opened up by von Neumann's famous Minimax Theorem. The theory of minimax is a new and important concept in science—and in life—as it is one of the few theories we have that defines how to proceed rationally in what were previously seen as irrational situations. So if nothing else, game theory has cut one of the major Gordian knots of philosophy: the barrier between rational and irrational action. It's hard to ask any more from a mathematical theory than that. Put succinctly as a paraphrase of a remark once made by numerical analyst Richard W. Hamming in the context of computation, "the purpose of game theory is insight, not solutions."

41

# CHAPTER

# 2

# The Brouwer Fixed-Point Theorem

*Topology*

# Needles and Haystacks

On November 13, 1993, the University of Notre Dame football team defeated Florida State University by the score 31 to 24 to claim the top spot in the weekly college football rating polls. Prior to this game, both the Associated Press (AP) poll of selected sportswriters and the *CNN/USA Today* poll of football coaches had ranked Florida State first. But following the Notre Dame game, Florida State was dropped to second place by the AP, while falling to third place behind the University of Nebraska in the coaches' poll.

Not surprisingly, the media voting reflected a fanatic desire to see a rematch between Florida State and Notre Dame in one of the New Year's Day bowl games, something that could occur only if Florida State were to remain in second place (a match-up that never did take place, since Notre Dame itself lost the very next weekend to a spunky team from Boston College). Such obvious maneuvering by the sportswriters touched off howls of outrage from a number of commentators and fans (especially those in Florida). And just to indicate how bad the situation is with the various polls, Figure 2.1 shows the rankings for the 1993 college season for games through November 20. The first four columns represent the ratings of the AP, *CNN/USA Today, The New York Times* computer poll, and the *USA Today* computer poll, respectively. The fifth column is the ranking by mathematician James Keener, based on ideas we'll discuss later in the chapter, while the final column is a ranking by statisticians that relates future performance to the outcomes of previous games.

Like Banquo's ghost, this sorry situation, in which factors seemingly unconnected with the intrinsic merits of the football teams themselves influence the teams' rankings, seems to haunt the college football scene every autumn. As a counter-reaction, a few obsessive mathematicians, statisticians, and sports aficionados have tried to concoct a clean mathematical answer to the messy human problem of measuring the power of football teams. We want to sketch the general idea of these ranking schemes, showing their direct connection to the issues of this chapter.

The most direct rating method is to assign a numerical ranking to each of the $N$ teams that is based on a team's games with other teams. The ranking should depend on both the outcome of the game(s) and the strength of the team's opponents. Suppose we let $r_1$ be the ranking of team 1. For the sake of exposition, let's say this is some number between

| Team | Record | 1 | 2 | 3 | 4 | 5 | 6 |
|---|---|---|---|---|---|---|---|
| Florida State | 10-1-0 | 1 | 2 | 1 | 1 | 1 | 1 |
| Nebraska | 10-0-0 | 2 | 1 | 2 | 4 | 6 | 11 |
| Auburn | 11-0-0 | 3 | — | 5 | 13 | 11 | 22 |
| Notre Dame | 10-1-0 | 4 | 4 | 9 | 3 | 4 | 4 |
| West Virginia | 10-0-0 | 5 | 3 | 7 | 12 | 9 | 19 |
| Tennessee | 8-1-1 | 6 | 5 | 4 | 2 | 2 | 2 |
| Florida | 9-1-0 | 7 | 6* | 3 | 5 | 5 | 6 |
| Texas A&M | 9-1-0 | 8 | 6* | 6 | 6 | 3 | 3 |
| Miami (Fla.) | 8-2-0 | 9 | 9 | 8 | 7 | 7 | 5 |
| Wisconsin | 8-1-1 | 10 | 8 | 16 | 22 | 8 | 21 |
| Boston College | 8-2-0 | 11 | 12 | 13 | 20 | — | — |
| Ohio State | 9-1-1 | 12 | 10 | 11 | 9 | 17 | 10 |
| North Carolina | 9-2-0 | 13 | 13 | 14 | 16 | 21 | 18 |
| Penn State | 8-2-0 | 14 | 11 | 15 | 11 | 13 | 8 |
| U.C.L.A. | 8-3-0 | 15 | 14 | 10 | 8 | 16 | 9 |
| Oklahoma | 8-2-0 | 16 | 15 | 12 | 19 | 15 | 13 |
| Alabama | 8-2-1 | 17 | 17 | 21 | 14 | 14 | 16 |
| Colorado | 7-3-1 | 18 | 18 | 18 | 17 | 25 | 25 |
| Arizona | 8-2-0 | 19 | 16 | 25* | 21 | 20 | 15 |
| Kansas State | 8-2-1 | 20 | 19 | 24 | — | — | — |
| Indiana | 8-3-0 | 21 | 21 | — | — | 18 | 20 |
| Virginia Tech | 8-3-0 | 22 | 20 | 19 | 25 | — | — |
| Michigan | 7-4-0 | 23 | 22 | 17 | 10 | 10 | 7 |
| Clemson | 8-3-0 | 24 | 23 | — | — | — | — |
| Michigan State | 6-3-0 | 25 | 24 | 25* | — | — | — |
| Southern California | 7-5-0 | — | 25 | 23 | 15 | 24 | 12 |

**Figure 2.1** Football rankings for the 1993 season (through November 20).

0 and 100. The quantity $r_1$ can be thought of as a point in the interval of real numbers between 0 and 100. If $r_2$ is the ranking of team 2, then we can represent the pair of rankings $(r_1, r_2)$ as a point in a square with sides of length 100. But we have $N$ teams, so we list their rankings as a vector

$$ r = \begin{pmatrix} r_1 \\ r_2 \\ \vdots \\ r_N \end{pmatrix}, $$

where the number $r_j$ is just the ranking of team $j$. This is a point in a square having $N$ sides, each of length 100. So the complete ranking vector $r$ belongs to the space of points consisting of this hard to visualize—but easy to mathematically write down—"hypersquare."

One way to measure the strength of team $i$ is to simply add up the results of that team's games with the other teams that it has played. This leads to the formula

$$r_i = (a_{i1}r_1 + a_{i2}r_2 + \cdots + a_{iN}r_N)/n_i, \quad i = 1, 2, \ldots, N, \qquad (*)$$

where the number $a_{ij}$ is a nonnegative quantity that depends on the outcome of the game between teams $i$ and $j$, and is constructed somehow to measure the relative strengths of the two teams in that game. Here we divide $r_i$ by the quantity $n_i$, the total the number of games played by team $i$, in order to prevent teams from accumulating a large score simply by playing more games. One plausible—although not very useful—way to assign the number $a_{ij}$ is to take it to be 1 if team $i$ won the game with team $j$, $\frac{1}{2}$ if the game ended in a tie and 0 if team $i$ lost the game. We'll come back to the problem of how to *best* choose the number $a_{ij}$ at the end of the chapter.

However one defines the quantities $a_{ij}$, it's not unreasonable to suppose that the ranking of team $i$ given by Eq. $(*)$ above is directly proportional to its strength. After all, that's what a team's ranking is supposed to measure. To streamline our notation, let's define the number $\alpha_{ij} = a_{ij}/n_i$ to be the entry in row $i$ and column $j$ of a matrix $A$, so that

$$A = \begin{pmatrix} \alpha_{11} & \alpha_{12} & \cdots & \alpha_{1N} \\ \alpha_{21} & \alpha_{22} & \cdots & \alpha_{2N} \\ \vdots & \vdots & \ddots & \vdots \\ \alpha_{N1} & \alpha_{N2} & \cdots & \alpha_{NN} \end{pmatrix}.$$

Then by the rules of matrix multiplication

$$Ar = \begin{pmatrix} (a_{11}r_1 + a_{12}r_2 + \cdots + a_{1N}r_N)/n_1 \\ (a_{21}r_1 + a_{22}r_2 + \cdots + a_{2N}r_N)/n_2 \\ \vdots \\ (a_{N1}r_1 + a_{N2}r_2 + \cdots + a_{NN}r_N)/n_N \end{pmatrix}.$$

But this is just Eq. $(*)$, which means we can write this equation more compactly as $r = Ar$. This tells us that the ranking vector $r$ is a solution of the equation $Ar = r$. So the ranking vector we seek is simply a point in the hypersquare of vectors that is left unchanged when it is multiplied by the matrix $A$; it is what's called a *fixed point* of $A$. Here's a slightly more algebraic way of looking at the situation.

Suppose we have a ranking vector $r^{(0)}$ that represents an estimate of the relative strengths of the teams at the beginning of the season when no games have yet been played. This ranking may come from the opinions of sportswriters, fans, coaches, or whatever. Now after the first week's games are over, this ranking vector changes. It is transformed to a new point $r^{(1)}$ in the hypersquare in accordance with the rule

$$r^{(1)} = Ar^{(0)}.$$

Thus, $r^{(1)}$ is the ranking after week 1. Similarly, after the games of week 2 the ranking vector is

$$r^{(2)} = Ar^{(1)} = A(Ar^{(0)}),$$

and so on. In this way we see that each week the ranking vector moves from one point in the ranking hypersquare to another.

Now suppose it happened in week 5 that the ranking vector turned out to be the same as in week 4. This would mean that

$$r^{(5)} = Ar^{(4)} = r^{(4)}. \qquad (**)$$

But now consider the situation as regards the ranking for week 6. By definition, we have

$$r^{(6)} = Ar^{(5)}.$$

But from Eq. (**) we find that

$$r^{(6)} = Ar^{(5)} = Ar^{(4)} = r^{(5)}.$$

Thus we see that if the ranking vector stays the same for 2 consecutive weeks, it will remain at that point for all future weeks. So the process of moving points about in our hypersquare from week to week to get an updated ranking grinds to a halt. In this case, the "final" ranking from such a procedure is simply a point in the hypersquare that remains fixed (invariant) under the transformation $A$.

We'll return to the actual determination of such a fixed point later on. For now, it suffices just to note that solving this very down-to-earth (although admittedly not very earth-shattering) problem of ranking football teams ultimately comes down to the solution of an equation. We want to find the vector $r$ such that $r = Ar$, or what's the same thing, we seek the solution of the equation $r - Ar = 0$.

Fixed-point theory is the most general tool in our mathematical arsenal for telling us whether or not a given equation has a solution of a specified type. For instance, we'll see later that the set of all paths from the Earth to the moon is a curve in a certain space whose elements ("points") are curves. In that case, we want to know if there is some point in this space, that is, a path, that will take us from one point to another at a certain fixed cost in energy. The way this question is usually answered is to guess such a path, and then successively transform this guess until we find a transformed curve that is the same as its predecessor. That is, until the current curve remains fixed under the transformation process, just as the vector $r^{(5)}$ remained fixed under the transformation $A$ in the football-ranking example just considered. Such a final curve is a fixed point of the transformation, and under suitable conditions will be a solution to the original minimal-energy problem.

Unfortunately, most of the classical theory we'll discuss here does not tell us how to actually *find* a fixed point. But if you are asked to look for a needle in a haystack, it's nice to know that there really is a needle in there before you get out your pitchfork and start digging around. It's this question of existence of a solution that the Brouwer Fixed-Point Theorem is designed to answer. Now let's talk for a while about surfaces and spaces. Then we'll return to discuss how these considerations relate to finding solutions of equations.

At first glance, one may think that there is no connection that matters between the problem of determining the existence of the solution to an equation and that of determining why, for example, a geometrical space like an American football is different from another space like a doughnut. But that's topology for you, always turning up unexpected connections between things that look as if they ought to be unrelated.

## The Shape of Space

It is reported that over the entrance to Plato's Academy in ancient Athens was inscribed the dictum, "Let no one ignorant of geometry enter here." And to this day, most people seem to find geometry the most congenial of the three main divisions of mathematics—algebra, geometry, and analysis—probably because it is the most visual and therefore easiest to follow. In broad terms, geometry is the study of the spatial properties of objects. We are all familiar with some of these objects from everyday life, things like a line on a piece of paper or a circle drawn on the surface of a balloon. Simple objects like flower vases and the knots in your shoelaces are also commonplace examples of geometrical objects. Since geometry is the mathematical idealization of space, a natural way to organize its study is by dimension. First we have points, objects of dimension 0. Then come lines and curves, which are one-dimensional objects, followed by two-dimensional surfaces, and so on. A collection of such objects from a given dimension forms what mathematicians call a "space." And if there is some notion enabling us to say when two objects are "nearby" in such a space, then it's called a *topological space.*

Mathematical spaces come in many different—and sometimes quite exotic—flavors, in which the "points" of the space might be things like complex numbers, infinitely differentiable functions, or even more complicated mathematical gadgets. This is where the power of abstraction comes into play. And there is no area of mathematics where thinking abstractly has paid more handsome dividends than in *topology,* the study of those properties of geometrical objects that remain unchanged when we deform or distort them in a continuous fashion without tearing, cutting, or breaking them. A good example of such a topological distortion is seen in Figure 2.2, which shows how we can continuously transform a coffee cup into a doughnut by pushing the top rim and bottom surface of the cup together, thereby removing the volume that normally holds the coffee.

Usually when we talk about geometry we have in mind a set of objects, or "shapes," belonging to some space, together with various properties of those shapes and/or relationships among them. The best example is the plain old plane geometry of Euclid that we all labored over in high school, doing our best to tease out some of the logically airtight secrets of the properties of simple shapes like straight lines, circles, trapezoids, and all the rest. One such secret, for example, is the familiar fact that the sum of the angles of a triangle equals 180 degrees. But we also know that these secrets depend upon the underlying world in which the objects "live." So, for instance, Euclid's objects live in

**Figure 2.2**  Topologically transforming a coffee cup into a doughnut.

the everyday world of the flat two-dimensional plane, a piece of paper if you like. But if we regard our lines, trapezoids, and circles as living in quite another world, say on the surface of a sphere, then we get a very different set of properties, such as the fact that the sum of the angles of a spherical triangle is always *greater* than 180 degrees. So the character of the underlying space in which the objects are embedded may affect the geometrical properties of the objects.

Through the middle part of the nineteenth century, many different geometries—euclidean, noneuclidean, analytic, affine, projective, and so forth—were introduced and studied. But it was not until Felix Klein published his famed Erlanger Program in 1872 that a coherent framework for the systematization of geometry was finally mapped out. Klein's program for the study of geometry was based on the key idea that what separates one geometry from another is how the objects of study behave when subjected to particular types of transformations. Klein saw that the crucial element in this research program was to uncover those propositions about the objects that remain fixed, or invariant, under a given set of transformations.

For example, euclidean geometry is distinguished from other geometries by properties that remain unchanged when objects in the plane like circles and triangles are subjected to *rigid motions,* that is, rotation through a fixed angle, translation to a new location in the plane, and/or reflection in a fixed line. These are transformations that do not distort or "warp" the object, unlike continuous deformations of the coffee-cup variety discussed above. Since properties such as the angles of a triangle and the area of a circle all ultimately reduce to the invariance of the length of a line in the plane when the line is transformed by a rigid motion, what singles out the geometry of Euclid from all other geometries is that all propositions about the objects remain true if we transform the objects by picking them up and moving them to a new position, rotate them, or reflect them in a fixed line—but without twisting, stretching or distorting them in any other way.

The set of all rigid motions forms what mathematicians call a *group* of transformations, and Klein's Erlanger Program regards every subdiscipline of geometry as dealing with those properties of geometrical objects that remain invariant under a particular such group. What Klein was arguing is that it's the *group* of transformations that gives a geometry its "individuality," not the objects, and that we should focus our attention on how things transform under the elements of some group instead of looking at the things themselves.

One of our principal goals in this chapter is to follow Klein's dictum—but only insofar as it pertains to surfaces. Therefore we will start with two surfaces and ask: Can one of these surfaces be continuously deformed into the other? That is, are these two spaces topologically equivalent? After all, it's a lot easier to do calculations with nice round spheres like a basketball than with odd-shaped objects like a rugby ball. But if you know how to transform a basketball into a rugby ball by using some group of transformations, then you can just do your computations in the world of spheres, transforming the final results back to the rugby-ball situation at the end. It's the topological equivalence of the two spaces—the basketball and the rugby ball—that makes this possible. And it's topology that tells us when two different-looking objects, or spaces, really can be transformed one to the other.

We saw above that if the surfaces are those forming a coffee cup and a doughnut, the answer to our question is yes, they are topologically equivalent surfaces. But we're after bigger game. We want a general theory that will answer the question for all pairs of surfaces. Development of such a theory involves looking for properties of a surface that remain invariant under any continuous transformation of the surface. If the numerical values of these properties for one surface are the same for another surface, then the two surfaces *might* be equivalent, that is, continuously transformable one to the other. But if the numerical value of even one such property is *not* the same for the two surfaces, then they are definitely not equivalent. So we need to ask about the kinds of properties and how many of them we need in order to settle the question of when two surfaces are topological equivalent.

As we'll see, it's possible to completely answer these questions for most surfaces. But things get a lot trickier for higher-dimensional spaces, even when we move just from surfaces to ordinary three-dimensional space. This is why the chapter is confined mostly to a discussion of surfaces (besides the fact that we can draw nice pictures in that case). But what's the connection here with the fixed points of a transformation?

Suppose instead of two different surfaces, $X$ and $Y$, we consider the case when $X$ and $Y$ are higher-dimensional topological spaces. And let's take the special case when $X = Y$. So we are considering continuous transformations of a space to itself. We can now ask a similar kind of question: What, if anything, remains invariant under all such transformations? This is where fixed points enter the scene, since there may be *no points* that remain fixed. For example, think of the rotation of a circle by a fixed amount. In this case, every point moves to some other point under the rotation; hence, there are no fixed points. So the matter of the existence of fixed points becomes that of determining the properties a space $X$ must possess so as to ensure that for every continuous (that is, topological) mapping of the space to itself there is at least one point of the space that remains fixed.

It's mathematically interesting if a space has a fixed point under a certain class of transformations. But it's also interesting if it doesn't. Practically speaking, the two questions are of equal interest and importance. As the football-ranking and Earth-to-moon examples show, the set of possible solutions to many problems can be regarded as points in an abstract mathematical space. But whether any of these points actually solve the problem at hand can almost always be formulated as a question about whether there exists a point in the space that stays put when we subject the space to a particular type of transformation. And a continuous transformation of a space is a *very* general type of transformation, one that essentially allows any kind of deformation of the space that doesn't "tear" it into pieces. This generality allows us to express many questions concerning the existence of solutions to equations arising in biology, economics, physics, and engineering as fixed-point problems. Much of the latter half of the chapter is devoted to showing how this can be done.

So we find that the common link between the problem of classifying spaces (for instance, surfaces) and the solving of equations (that is, finding fixed points) is the notion of a continuous transformation of a space $X$. In one case, $X$ is mapped continuously to a possibly different space $Y$; in the other case, $X$ is mapped to itself. But the nature of the question we want to answer remains the same in both cases. We want to find a point or a property of the space that remains unchanged when we subject the space to a continuous transformation, and then use that property or point as a way of determining (1) if the spaces $X$ and $Y$ are topologically equivalent, or (2) if the space $X$ has a fixed point. Finding those invariant properties or points is a large part of what topology is all about.

# Topology

OK, now we know that topology is the study of those properties of geometrical objects that stay the same when the objects are subjected to a *continuous* transformation. Fine. So instead of restricting our set of transformations to rigid motions that define the properties we call euclidean geometry, let's now admit any transformation or mapping of the objects, provided only that two points that start close together also end up close together. But we do not want to allow *any* continuous transformation of the space, but only those continuous transformations such that every point of the transformed space corresponds to one and only one point of the original space, and conversely. In other words, we consider transformations that are continuous, have a continuous inverse, and are "onto," meaning that every point in the target space corresponds to some point in the source space of the transformation. Technically, such transformations are called *homeomorphisms*. So transformations that tear or break are not allowed, but we welcome those that bend and stretch.

The kinds of transformations allowed in topology are so general that only the most basic geometrical properties of an object remain unchanged when we deform the object by choosing from among the whole spectrum of possible continuous transformations. For this reason, topology is often called "rubber-sheet geometry." This reflects the idea that we can draw objects on something like the surface of a balloon or a surgeon's glove, and the only properties of the object that do not change when we stretch, shrink, or in some other way continuously deform the rubber sheet are those that we term *topological.*

It's evident, I think, that topological properties are quite different from the invariant properties we're familiar with from euclidean geometry. For example, the property of being a triangle is not a topological one, since we can continuously deform a triangle into a circle. Similarly, the property of being an American football is also not topological, because we can continuously squeeze the football into a baseball—or, for that matter, into a European football (a soccer ball).

So what kinds of properties are topological? One such property is that of having a hole. Thus, while a doughnut can be continuously transformed into a coffee cup, it cannot be continuously twisted and molded into a basketball, an object without a hole. Having an edge (that is, a boundary) is another topological property. Thus, a circle has no edge, while a straight line does (its two endpoints). And there is no way to continuously deform the circle into the line, since any such transformation would have to map the two endpoints of the line onto the same point of the circle, thereby violating the requirement that the transformation be one-to-one. In other words, we would have to break the circle to map it to the line.

For many applications, both inside and outside mathematics, it's of considerable importance to know what kinds of geometrical objects can be continuously transformed, one to the other. This is because some kinds of standard geometrical objects like triangles and circles and spheres are easy to study mathematically and to do calculations with, while others—like American footballs—are not. So if we have at our disposal a transformation that continuously deforms a "hard" space like a football into an "easy" space like a sphere, we can do all our mathematical investigations on the sphere, transferring these results later to the football using this known transformation. So to simplify our mathematical lives, we would like to know all those objects that can be deformed, say, into a sphere. Then we can lump all such objects into the same category, call it SPHERES, and investigate the properties of every element in this set in one fell swoop just by studying a "standard" sphere. The set SPHERES is what's called an *equivalence class,* while the standard sphere (the one that looks like a basketball) is called a *canonical representative* from the class SPHERES. Since, by definition, any two objects in the class SPHERES are deformable one to the other via a topological (that is, continuous) transformation, we call them *topologically equivalent.* Let's look a bit further into this idea.

## Topological Equivalence

For simplicity, we confine our attention to topological spaces that are bounded surfaces. Roughly speaking, a *surface* is any topological space that looks like an ordinary two-dimensional plane—*locally.* This means that an ant walking around in such a space would not see the *global* twists and turns of the space, but would think that it was living on the surface of a flat piece of paper, just as we humans think we're living on a flat plane until we get into a high-flying airplane like the Concorde or a space shuttle and actually see the curvature of the Earth.

Some surfaces like the sphere are bounded, while others like the plane $R^2$ are not. You can walk around on a sphere forever and never leave the surface, just as most of us spend our lives walking around on the surface of the Earth. Moreover, some surfaces, like a closed disk in the plane, have boundaries, while others like the sphere have no boundary at all. This means that for the disk there is a curve that "fences in" its points, and if you cross this curve you can leave the surface. Such a curve is called the *boundary* of the surface. There are no such closed curves on the surface of the sphere; if you walk across any closed curve on the sphere, you're still on the sphere. So the sphere is bounded and closed—but without boundary. Connected surfaces (those all in one piece) having no edge are called *closed surfaces*. So a sphere is closed while a disk is not.

Figure 2.3 shows a number of surfaces, some of which are topologically equivalent. For instance, we have seen that it's possible to continuously deform the doughnut-shaped object technically termed a *torus* into a sphere with one "handle" by grabbing a part of the torus and pulling on it hard enough. It's less transparent—nevertheless true—that the trefoil is also topologically equivalent to the torus, hence to a sphere with one handle. The trefoil, incidentally, shows that the way a surface is twisted or knotted is not an intrinsic topological property of the surface itself, but rather depends on how the surface has been embedded into the surrounding space. So if you were a two-dimensional cockroach living on the surface of the trefoil, it would be impossible for you to tell that you were not living on the surface of a torus instead. You would have to jump outside your twisted, folded-over trefoil world to discover that it's knotted. So being knotted ("naughtiness"?) is not a property preserved under a topological transformation. But having a hole ("holiness"?) is.

A principal goal of topology is to try to classify all topological spaces. For closed surfaces, this problem reduces to providing a listing of closed surfaces so that every closed surface is continuously transformable to one and only one surface on the list. Such a list can actually be given explicitly for such surfaces—and it turns out that it is a very short one. In fact, the list contains just two fundamentally different types of closed surfaces: spheres with handles and spheres with something called *cross-caps,* which we'll define in a moment. Every other closed surface can be deformed continuously into one of these two types. But to see this, we first need to stir a few more mathematical ingredients into our topological stew.

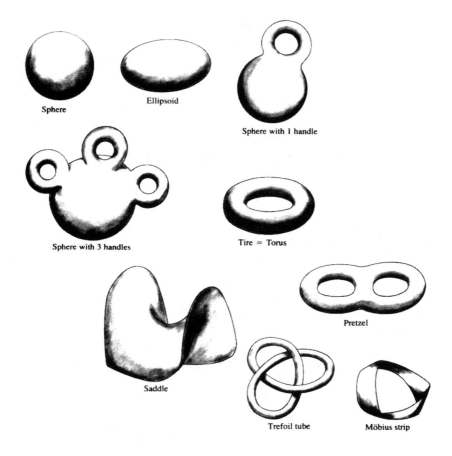

**Figure 2.3** Some surfaces.

Let's consider what is perhaps the most well chronicled, if not the most important, of all surfaces, the *Möbius band*. This gadget is formed by taking a strip of paper, giving one end a half twist and pasting the resulting ends together. You end up with an *open* surface having one edge, rather than a closed surface with no edge. If you were able to walk about on the surface of the Möbius band, one complete tour of the territory would look something like what's shown in Figure 2.4. Strangely enough, unlike such a tour on the surface of a sphere in which you would just come back to where you started as if nothing had happened, a Möbius man or woman does not come back to the starting point of his or her tour with the same orientation. Instead, the tour transforms the

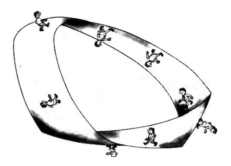

**Figure 2.4**   A tour of the Möbius band.

Möbius person into his or her own opposite, that is, he or she comes back "upside down." Surfaces having this property are called *nonorientable,* because it's impossible to consistently define the notions of left- and right-handedness on them. On the other hand, a surface like the sphere is *orientable,* since if you go around it on, say, a great circle, you come back in the same orientation as when you began.

Returning now to the problem of classifying all closed surfaces, it turns out that all closed, orientable surfaces can be continuously transformed into a sphere with a certain number of handles. The number of handles *g* is called the *genus* of the surface. So, for instance, the sphere is topologically equivalent to—a sphere, an object with no handles. A torus is deformable to a sphere with one handle, as we saw earlier, while Figure 2.3 showed that a pretzel can be molded into a sphere with two handles, and so on. Thus, from a topological point of view the only orientable surfaces we need concern ourselves with are spheres with handles. But what about nonorientable surfaces?

The situation for nonorientable closed surfaces is just a little bit trickier, geometrically speaking, than that for orientable ones. To see what the standard representatives are for this sort of surface, we need to introduce a new idea. Take any surface you like and cut a hole in it. This hole has a single, circular edge. The Möbius band also has a single, circular edge. So we can sew the Möbius band across the hole edge to edge, as shown in Figure 2.5. This can only be visualized in three dimensions if the Möbius band is allowed to intersect itself; in mathematical reality, there is no such intersection. The surface we end up with after sewing-in the Möbius band is called a *cross-cap.* Since

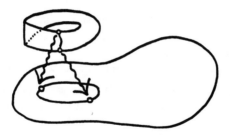

**Figure 2.5**   The cross-cap.

it would take us too far afield to go into the details and properties of
this object, let me refer the interested reader to the volumes cited in the
bibliography for additional information. Suffice it to say here that the
cross-cap is topologically equivalent to the Möbius band; hence, it is
nonorientable. It can be shown that any nonorientable closed surface
is topologically equivalent to a sphere with a certain number $g \geq 1$ of
cross-caps. As in the orientable case, the number $g$ is called the genus
of the surface.

    We now summarize what we have just learned about topological
equivalence for closed surfaces in the following fundamental result.

---

**CLASSIFICATION THEOREM FOR CLOSED SURFACES**

   *a. Every orientable closed surface is topologically equivalent to a*
      *sphere with a certain number (g $\geq$ 0) of handles.*
   *b. Every nonorientable closed surface is topologically equivalent to*
      *a sphere with a certain number (g $\geq$ 1) of cross-caps.*   ■

    The above considerations lead to the somewhat comforting conclu-
sion that the list of standard surfaces to which any other closed surface
is topologically equivalent, contains only spheres with either handles
(for orientable surfaces) or cross-caps (for nonorientable surfaces). Unfortunately, the situation quickly becomes more complicated when we
pass to higher-dimensional topological spaces. Nevertheless, the general
program is the same: Produce a list of "standard" topological spaces of
the requisite dimension, such that any space of the same dimension can
be continuously transformed to exactly one element on the list. A sphere
with handles or cross-caps exhausts the list of standard closed surfaces;

in higher dimensions, only partial classifications are available involving more complicated objects than just higher-dimensional analogs of spheres with handles or cross-caps.

It may have occurred to the perceptive reader that one could try constructing a surface that contains *both* a handle and a cross-cap. This could be done, for instance, by cutting three holes from a sphere, then attaching the two ends of a cylinder to two of the holes, and sewing a Möbius band to the third hole. These operations would seem to generate a surface that is simultaneously orientable (since it has a handle) and nonorientable (since it has a cross-cap). So which is it?

To resolve this conundrum, here's a short argument attributed to Christopher Zeeman (by way of Ian Stewart) showing that what you end up with is a nonorientable surface with *three* cross-caps. Think of a torus as a handle attached to sphere. Shrink the attaching disks down to small size. Then transport one of them around the Möbius band, in order to reverse its orientation, and then take it back to its original location. The torus now becomes a sew-in Klein bottle, which is two Möbius bands sewn together along their edges. Thus, the resulting surface has three Möbius bands sewn-in: the two from the Klein bottle, plus the one that was originally sewn-in to form the cross-cap.

What this small exercise shows is how just one, small, nonorientable bit of a surface makes the whole thing nonorientable. Basically, orientability is a global property. The original surface with a handle and a cross-cap only *appears* to be "locally orientable" over a big subset. But in topology "big" has no meaning, and we could easily deform that surface so that it has a "big" Möbius band and a "tiny" torus.

Just to see how tricky things become even when we move up just a single dimension, let's consider briefly one of the most famous outstanding problems in mathematics, the celebrated *Poincaré Conjecture.* This long-standing assertion involves what's called a simply connected, closed, three-dimensional manifold. This is just mathematical jargon for a space that looks like ordinary three-dimensional space near each of its points, and has no "edge", which here means that there is no two-dimensional boundary. Finally, "simply connected" means that the manifold is all in one piece, rather than being formed as the union of a collection of separate "chunks" of 3-space. The Poincaré Conjecture asserts that every such object is topologically equivalent to the 3-sphere.

A positive solution to the Poincaré Conjecture would lead to a complete classification of closed three-dimensional manifolds, similar to the one we gave above for closed, two-dimensional manifolds, that is, closed surfaces. But despite many attempts to prove this simple-looking statement, it remains one of the great unsolved problems of mathematics. Interestingly enough, though, the problem has been solved in *all* dimensions other than 3! So it seems that three-dimensional space is a bit like Goldilocks's porridge, lying between the "cold" dimensions 1 and 2, where there is not enough room for things to go too far wrong, and the "hot" dimensions greater than 3, where there is enough room to maneuver around obstacles standing in the way of a complete classification. So 3-space is not too hot and not too cold, but just right for the Poincaré Conjecture to remain open.

## The Fixed-Point Game

Let's play the simple paper-and-pencil game shown in Figure 2.6. The rules are as follows: First, place one dot anywhere on each of the two vertical lines $x = 0$ and $x = 1$. Now, without letting your pencil leave the paper, draw a curve $C(x)$ connecting the two dots that never touches the diagonal line dividing the square. It doesn't take much experimentation to see that this isn't much of a game, because there is no winning strategy. *Every* continuous curve connecting the two dots must necessarily touch the diagonal in at least one point. Here's why.

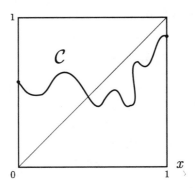

**Figure 2.6** The Fixed-Point Game.

61

Any continuous curve $C(x)$ starting on the vertical lines $x = 0$ and ending on the line $x = 1$ can be regarded as a transformation of the unit interval $0 \leq x \leq 1$ to itself, since it associates every point $x$ in this interval with another such point $C(x)$ that also lies in the range between 0 and 1. The diagonal line corresponds to points that map to themselves under this transformation; they remain fixed. Consequently, any point $x$ that maps to a point on the diagonal is a fixed point of the transformation $C$. In other words, it is a point for which $C(x) = x$. By arguments originally due to the Dutch topologist L. E. J. Brouwer, it turns out that every such curve mapping the unit interval to itself must possess at least one such point. So there is no way to win in the Fixed-Point Game.

Since the impossibility argument just given for why you can't ever hope to win in the Fixed-Point Game works for any and all continuous transformations of the unit interval to itself, the property of having a fixed point really seems to be more a property of the set of points constituting the unit interval than it is a feature of the particular transformation $C$. This observation leads us to ask what is it about sets like the unit interval that endow them with this "fixed-point property"? The answer to this question will enable us to tell *in advance* whether a given space has a fixed point when mapped to itself under a continuous transformation. And this knowledge, together with information about what spaces are equivalent to what other spaces, will then permit us to state with confidence when equations have solutions and when they don't—without having to ever actually explicitly display a solution. We'll know that the needle really is in the haystack before we invest time, energy, and money in trying to find it. Supplying ourselves with this kind of mathematical ammunition is the leitmotif of this chapter.

## Solving Equations

So far, we have focused our attention on the problem of determining when two spaces (that is, sets of geometrical objects) are the same, at least insofar as their topological properties go. Following in the steps of Felix Klein's Erlanger Program, we have approached this question by trying to uncover those properties of the objects that remain invariant under a particular group of transformations, the set of all continuous transformations of the space. But as we have seen above, there is another type of invariance question that is also of great importance in mathematics involving the determination of whether or not a given equation, like the one for the football ranking vector, has a solution. We want to show now how this general question can be mathematically reduced to the problem of determining whether a given continuous transformation of a topological space to itself has a fixed point. So we will now leave the theory of surfaces behind, devoting the remainder of the chapter to this key issue.

In discussing the Fixed-Point Game earlier, we were concerned with finding a fixed point of the transformation $C(x)$. It turns out, however, that is the same thing as finding a solution of the equation $G(x) = 0$. To see this, define the function $G(x)$ to be $G(x) = C(x) - x$. This argument goes in the opposite direction, too. Suppose we want to solve the equation $G(x) = 0$. Then if we introduce the identity operator $I$, whose effect is to do nothing, that is, $I(x) = x$ for every point $x$, we can simply subtract $G$ from this operator to obtain the new transformation $H = I - G$. The original equation $G(x) = 0$ then takes the form $G(x) = (I - H)(x) = 0$, or $H(x) = I(x) = x$. In other words, the solution we seek is a fixed point of the transformation $H$.

The above calculation shows that determining fixed points is the same as solving equations. This, in turn, is the raison d'être for much of mathematics. Consequently, it would be of great value to have some general results telling us under what circumstances a fixed point exists for a given transformation. This is exactly the content of the famous fixed-point theorems of Brouwer and others that we will consider over the next few pages. But before looking into these matters in greater detail, let's temporarily set the mathematics aside and show that the notion of a fixed point is not just an idle mathematical curiosity but impacts the world of real people and big money in a very important way.

## Dollars and Sense

In 1776, Scottish economist Adam Smith advanced the notion of an "invisible hand" guiding the economic process, thereby providing a rationale for how the selfish acts of greedy individuals can actually serve the interests of society as a whole. In 1874, almost exactly 100 years later, Leon Walras made the first concerted effort to formalize mathematically Adam Smith's view of a price as being this invisible hand, or equilibrating factor, serving to balance out supply and demand. Some years later, in 1932, John von Neumann gave a seminar in Princeton on his theory of economic processes, which was published in 1938 under the title, "Über ein Ökonomisches Gleichungssystem und eine Verallgemeinerung des Brouwerschen Fixpunktsatzes" ("On a System of Economic Equations and a Generalization of the Brouwer Fixed-Point Theorem"). In this theory, von Neumann established the existence of the best techniques of production to achieve maximum outputs of all goods at the lowest possible prices, with the outputs growing at the highest possible rates. Thus, at a single blow he transformed general equilibrium theory from statics to dynamics, using as his main analytical tool—the Brouwer Fixed-Point Theorem. Von Neumann's work is the direct forerunner of modern work by Yves Balasko, Stephen Smale, Richard Goodwin, and others on the dynamics of economic processes, especially on the ways in which economies can slip into various types of both cyclic and chaotic behaviors.

So successful was the mixture of mathematics and economics stemming from this work, that a century later this line of investigation received the ultimate legitimization when Kenneth Arrow was awarded the 1977 Nobel Prize in economics, mostly for work on what has come to be termed "general equilibrium theory." To add a bit of frosting to the cake, the Nobel committee honored Gerard Debreu in 1983 for further developments of general equilibrium theory reported in his magnum opus *Theory of Value,* a volume of such dazzling mathematical pyrotechnics that it turns completely upside down the traditional view of mathematicians as being the keepers of the abstract processes, that dubious honor having now clearly passed on to the economists. Here is a simple example of equilibrium economics, showing how the ideas of supply, demand, and prices relate to the existence of fixed points.

## Guns and Butter

Suppose we have an economy consisting of two consumers who trade just two commodities: guns and butter. Assuming there is no production, the consumers would like to determine prices for these two commodities such that at these prices the market supply equals the demand. For the sake of definiteness, let's assume the quantities of the two commodities owned by the consumers prior to any trades are given by the table:

|  | Guns | Butter |
|---|---|---|
| **Consumer 1** | 8 | 0 |
| **Consumer 2** | 4 | 8 |
| *Totals* | 12 | 8 |

Now suppose the prices of guns and butter are $p_G = \$1$ and $p_B = \$2$, respectively. Then Consumer 1's income, obtained by selling guns and butter at market prices, will be $8 \times \$1 + 0 \times \$2 = \$8$, while Consumer 2 will have an income of $4 \times \$1 + 8 \times \$2 = \$20$. Of course, the two consumers will each have demands for a certain amount of guns and butter at the prevailing market prices. To make things simple, suppose one-fourth of Consumer 1's income is spent on guns and three-fourths is spent on butter, while Consumer 2's income is divided equally between the two commodities. General equilibrium theory seeks price levels for guns and butter at which the supply equals the demand.

With the arbitrarily chosen price levels $p_G = \$1$ and $p_B = \$2$, we have seen that Consumer 1 has an income of $8 and Consumer 2 receives $20. The demands for guns and butter will then be:

|  | Guns | Butter |
|---|---|---|
| **Consumer 1** | 2 | 3 |
| **Consumer 2** | 10 | 5 |
| *Totals* | 12 | 8 |

Voilà! At these price levels the supply and the demand for each commodity are equal. So the markets for guns and butter "clear," meaning that at these prices the economy is in equilibrium.

Of course, in this example we blithely plucked the prices out of thin air and then verified that the markets cleared at these levels. The goal of general equilibrium theory is first and foremost to establish the *existence* of such a set of equilibrium prices, and then to provide methods by which to calculate those magical price levels. The theory states that there always exists a set of prices at which supply equals demand for all goods, a result whose only known proof comes from showing that these prices are the fixed points of a particular transformation. This is a consequence of the fact that we can regard the prices as the elements in a vector, each of whose entries is a nonnegative real number. The set of all such vectors having $n$ elements constitutes a topological space, and

65

under reasonable conditions the price-setting mechanism in the economy is a continuous transformation of that space to itself, that is, it moves prices from one point in the space of prices to another. So let's look at how fixed-point theory allows us to establish these equilibrium price levels for a simple three-goods economy. We'll then come back to the general case of an economy with more than three goods.

Our economy is one in which three goods—guns, butter, and cars—are traded at prices $p_G$, $p_B$, and $p_C$, respectively. These prices generate a supply and demand for the goods, just as we saw a moment ago for the guns-and-butter economy. For brevity of notation, we will write the prices as the *price vector* $p = (p_G, p_B, p_C)$. In general, a particular price level does not lead to a market equilibrium, which means that there is an excess demand for one or more of the commodities. The economy will be in equilibrium for a set of prices at which the supply of each good is at least as great as the demand, and such that there is no excess demand.

The first mathematical trick is to recognize that it is only the relative price levels that count. So we can normalize prices. All this means is that if we multiply each price by a fixed amount by, for example, converting prices given in U.S. dollars to prices expressed in Japanese yen, nothing has really changed. So for mathematical convenience let's agree to quote the prices in a "currency" such that every good is priced between 0 and 1, and that the sum of the prices of all goods in the economy is 1. The way to form these relative prices is to divide each absolute price by the sum of all the absolute prices. Then each relative price vector $p$ in a three-goods economy is a point of the price triangle $P$ shown in Figure 2.7. With this bit of mathematical legerdemain, we can formulate the problem of the existence of an equilibrium price vector in the guns-butter-cars economy as a fixed-point problem. Here's how.

Each set of prices is a point $p$ inside or on the price triangle $P$. As the demand and supply changes for the three goods, so do the prices, with goods having an excess demand commanding a higher price than those in oversupply. Suppose the initial price levels for the three goods are $p_G^0$, $p_B^0$, and $p_C^0$, respectively. Now prices change due to the supply/demand imbalance at these prices. This means that we have a shift from the point $p^0 = (p_G^0, p_B^0, p_C^0)$ in the price triangle $P$ to a point $p^1 = (p_G^1, p_B^1, p_C^1)$. But we can think of the change $p^0 \rightarrow p^1$ as the result of a transformation $T$ that moves points in the price triangle to other points in the triangle. In economic terms, such a trans-

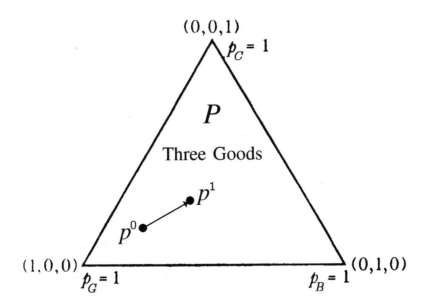

**Figure 2.7** The price triangle for a three-goods economy.

formation can be regarded as a price-setting mechanism, and what we need for economic equilibrium is the existence of a point $p$ in the price triangle such that $T(p) = p$. In other words, we need a fixed point of the transformation $T$.

For our purposes, all we demand of the price-setting mechanism $T$ is that it transform a price to a price (that is, that it does not take points in the price triangle $P$ outside the triangle), and that it be continuous. Roughly speaking, this latter requirement means that if two sets of prices $T(p)$ and $T(p')$ are "close," then the previous prices $p$ and $p'$ must also have been close. It turns out that there are a lot of price-setting procedures that both mathematically and economically satisfy these very weak conditions.

By the continuity of $T$, together with geometric properties of the price triangle—properties that will be outlined in a moment—the famous Brouwer Fixed-Point Theorem assures us that there always exists at least one set of prices that is a fixed point of the price-setting scheme $T$. The reader will have no problem seeing how to extend this setup to an economy in which there are an arbitrary, but finite, number of goods being traded, rather than just the cars, guns, and butter we have been trading here. Thus, the fixed-point results establish the vitally important fact that in a pure exchange economy, there exist prices at which all agents are satisfied, that is, at which the markets clear for all goods.

We now abandon the world of economics for the universe of mathematics, focusing our attention on the theory of fixed points and how one might go about actually determining them.

## Disks, Squares, and Fixed Points

In 1910, the Dutch mathematician L. E. J. Brouwer published an article titled "Über Abbildung von Mannigfaltigkeiten" ("On the Mapping of Manifolds"), a path-breaking paper laying the foundations for much of modern topology. In particular, in this article Brouwer presented the first proof for what is now called the *Brouwer Fixed-Point Theorem*. Before stating this result in more general terms, let's first look at a particular version of the theorem for a square in the plane.

Consider a square divided into a grid of squares like a chessboard, and a continuous mapping $f$ of this square to itself. If $p$ is a point of the square, we let $q$ denote the point to which $p$ is mapped by $f$, that is, $q = f(p)$. Now let's label each of the squares of the chessboard according to the following rule:

1. the square is labeled R if $f$ moves each point of the square closer to the right edge of the board;
2. the square is labeled L if the transformation $f$ moves each point of the square closer to the left edge of the board;
3. the square is labeled N if it is neither an R nor an L square.

By this labeling scheme, we see that no square in the extreme right column can be labeled R, since $f$ cannot map points of the board to points off the board. By the same token, no square in the extreme left column can be labeled L. Moreover, since no point can be moved both left and right at the same time, an L square cannot border an R square (since adjacent squares must have at least one point in common). As a result of this setup, a king can never cross from the left side of the board to the right without entering an N square. Similarly, a rook can move from the bottom row of the board to the top row by traversing only squares labeled N. This means that we can draw a continuous polygonal

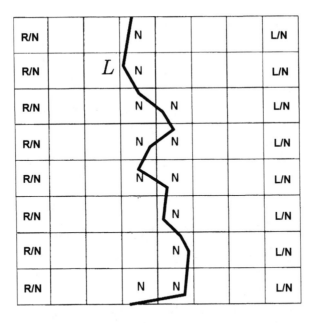

**Figure 2.8** A polygonal path through a labeled chessboard.

line $L$ from the bottom of the board to the top that passes only through squares labeled N. The overall situation is shown in Figure 2.8 above.

Now we take the line $L$ and draw an arrow from each point $p$ on the line to the corresponding point $q$ to which $p$ is mapped by the transformation $f$. At the starting point of the line on the bottom edge of the board, the arrow from $p$ to $q$ cannot be directed downward since it has to remain on the board. Thus, it must point in an upward direction (but not necessarily vertically). Similarly, the arrow from the termination point of the line on the top edge of the board must point downward. But since the mapping $f$ is continuous, as one moves along the line the directions of the arrows must change continuously. Consequently, there is a point $p_1$ on the line $L$ where the arrow from $p_1$ to $q_1$ is horizontal. But by the definition of an N square, the square containing the point $p_1$ must also contain a point $p_2$ whose arrow points vertically, straight up or straight down. This is because not all the arrows in an N square can point either to the right or to the left. So some arrows point right and some point left. The continuity of $f$ then ensures that at least one arrow must point straight up or straight down, that is, vertically. However, if

69

the square is sufficiently small, such a jump from the horizontal to the vertical in the direction of the square's arrows can happen only if the arrows themselves are small in length for all points of the square.

If we divide this chessboard into $n^2$ squares and make the division of the board finer and finer by letting $n$ go to infinity, the argument sketched above allows us to obtain a point $p^*$ for which the arrow from $p^*$ to $q^*$ vanishes. This means that $p^*$ is a fixed point of the transformation $f$. We will now take a more detailed look at the topological nature of Brouwer's results, along with some important extensions to more general situations.

## Compactness and Convexity

Since *every* continuous map of a square to itself has a fixed point, the fixed-point property cannot really be thought of as a property of the mapping but must somehow be inherent in the topological nature of the square itself. And by our earlier results on topological equivalence, whatever this property is it must be the same for a closed disk in the plane, since the square and the disk are topologically equivalent. To get at those special properties of the square and the disk that endow them with the fixed-point property, it's instructive to reexamine the situation in one dimension for the unit interval.

Recalling Figure 2.6, it's intuitively clear that if we consider the *half-open* interval formed by omitting the endpoint at $x = 1$, then there need be no fixed point. This is because our continuous curve $C(x)$ can get arbitrarily close to the line $x = 1$ without having to ever actually touch the diagonal line by "squeezing-in" just under the diagonal. So to make the fixed-point argument work, the underlying space should contain all the points of its boundary. Roughly speaking, any finite-dimensional topological space that is both bounded (like a fixed interval of the line) and contains all the points of its boundary, is what mathematicians call *compact*. But compactness of the space is still not enough to ensure the existence of a fixed point, as the following example shows.

Suppose we take a circle in the plane (not the disk of points bounded by the circle, but just the points forming the circle itself). This set of points is clearly compact, since the circle is a closed, bounded set in the plane. Now rotate the circle by any amount that is not a multiple

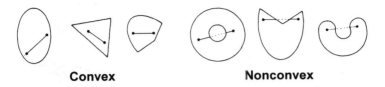

**Convex**     **Nonconvex**

**Figure 2.9**  Convex and nonconvex sets in the plane.

of 360 degrees. Such a rotation maps the circle to itself in a continuous fashion. Yet every point is mapped to a different point by such a rotation; hence, there is no fixed point.

It turns out that the topological property that's missing from the circle that prevents it from having the fixed-point property is that if we take any two points of the circle and join them by a straight line, there exist points on the line that are not on the circle. We say a topological space is *convex* if a straight line joining any two points of the space is contained entirely in the space. Figure 2.9 above shows some examples of both convex and nonconvex sets in the plane.

Compactness and convexity turn out to be all we need in order to state our main result on fixed points in a bit more formal terms:

---

**THE BROUWER FIXED-POINT THEOREM**  *Suppose* X *is a topological space that is both compact and convex, and let* f *be a continuous map of* X *to itself. Then* f *has a fixed point in* X, *that is, there exists a point* x* *in* X *such that* $f(x^*) = x^*$.  ■

Here's a simple example, sometimes called the "Porcupine Theorem," illustrating Brouwer's result in a somewhat whimsical fashion.

## Porcupines and Cyclones

If you look at the quills of a porcupine (not *too* closely!), you will find that they have a "parting" down the porcupine's back, and another along its stomach. Topologically, a porcupine is just a sphere (under the condition that it keeps its mouth shut and we neglect its internal structure), since all we need to do is shorten its legs and fatten it up somewhat to transform it

**Figure 2.10**  Porcupine quill patterns.

into a sphere. Now we can ask: Is it possible for the porcupine to lay all of its quills down flat? In other words, can we "comb" the quills of the porcupine so that all partings are eliminated? This would give a smooth prickly ball, with none of the arrangements of quills shown above in Figure 2.10.

With some effort, it can be shown using Brouwer's Fixed-Point Theorem that no perfectly smooth system of quills can exist. There must be at least one quill that stands out vertically at the center of a parting. The best that can be done is to comb the quills so that everything is smooth except at one point.

The basic idea of the proof of this extremely interesting fact about porcupines involves setting up a "combing map" that takes each of the porcupine's quills into the direction it points after being combed. With a little work, it can be shown that every such map is continuous. The Brouwer Fixed-Point Theorem is then used to conclude that there must be at least one quill whose direction remains unchanged under every such mapping, which then implies that it cannot be combed down flat.

The value of this result is pretty dubious, at least when it comes to porcupines. But if we consider the Earth's weather system, things come a bit closer to real-world concerns. The Earth is a sphere, topologically speaking, and it is of interest to ask about the direction of the winds that blow on its surface. In particular, we can inquire whether there is a point at which the winds are not blowing in any vertical direction. Such a spot would be the location of a cyclonic pattern in the winds. Regarding the flow lines of the wind as taking the place of the porcupine's quills, the "Porcupine Theorem" tells us that it is impossible to "comb" the wind so as to eliminate a point where the wind pattern is a cyclone. This is because the flow lines constitute a continuous mapping of the surface of the Earth to another point on the surface. (Technically: a map from the sphere $S^2$ to itself.) So, simply having knowledge of the topological class to which the Earth belongs (a sphere with no handles), we can say something interesting about wind patterns—without having any detailed knowledge of atmospheric physics, meteorology or, for that matter, cosmetology.

Oddly enough, it's possible to show that stable, smooth wind patterns *do* exist on planets shaped like a doughnut. The hole in the center moves such a toroidal planet into the topological class of a sphere with one handle. And Brouwer's results do not apply to such a space (why?). This fact, in turn, opens up the possibility for smooth wind patterns, smooth porcupines, and no cyclones.

Now that Brouwer's result assures us that such a thing as fixed points must exist, let's spend a few pages considering how one might go about actually finding them.

## Determination of Fixed Points

Brouwer established the fact that any continuous mapping of a compact, convex set to itself must have a fixed point. But his proof offered not so much as a hint as to how to go about finding such a distinguished point; the proof was an existential rather than a constructive result, made doubly odd by the fact that Brouwer spent much of the latter part of his career condemning the use of just such kinds of nonconstructive methods in the proof of mathematical propositions. But in a doctoral thesis written in 1928, just a few years after Brouwer's nonconstructive proof of the existence of fixed points, Emmanuel Sperner established a result now called *Sperner's Lemma* about the labeling of elements in a set of geometrical objects that has turned out to serve as the cornerstone of a whole family of algorithms for actually computing fixed points. Space constraints preclude a detailed discussion of these computational methods here, but we can spend a few moments to sketch the general idea in one dimension.

Consider the situation in which we have a continuous function $f$ mapping the unit interval to itself. Suppose we divide the interval into subintervals, labeling the ends of each subinterval according to the following rule: If the value of the function $f$ lies above the diagonal at an endpoint of the subinterval, then label that endpoint with a 1; but if the curve is below the diagonal at that endpoint, the endpoint is given the

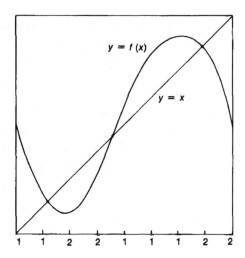

**Figure 2.11** A labeling of subintervals of the unit interval.

label 2. The general situation is shown in Figure 2.11 for a function $f$ having three fixed points in the unit interval. (Recall: The values of $x$ where the curve intersects the diagonal are the fixed points.)

The above labeling scheme has the property that each subinterval whose endpoints have different labels must contain a fixed point of $f$. This is because if the curve is below the diagonal at one end of the interval and above it at the other end, continuity demands that there must be a point in between where the curve intersects the diagonal. So if the subintervals are small enough, any point contained in a subinterval whose endpoints carry different labels serves as an approximate fixed point. And by letting the length of the subintervals shrink down to zero, the continuity of the mapping $f$, together with the compactness of the unit interval, ensures that any sequence of approximate fixed points will converge to a true fixed point of $f$.

It's easy enough to see in this one-dimensional setting that there must exist at least one subinterval whose endpoints are labeled differently. This is a one-dimensional version of Sperner's result. Sperner's Lemma extends this conclusion from subintervals of a line segment to the labeling of the vertices of a set of higher-dimensional counterparts of line intervals, things like triangles in two-dimensional spaces and tetrahedrons in 3-space. This fact, in turn, allows us to conclude that if we consider a continuous mapping of a compact, convex topological space to itself and partition the space into a large enough collection of such objects, there must be at least one element in the partition containing an approximate fixed point of the map.

By the 1960s, workers in the field of mathematical programming and economics had discovered the utility of Sperner's Lemma for approximating fixed points numerically. Computational algorithms have been produced that rapidly chop up the space of interest into small chunks, and which then search efficiently for the elements whose labels all differ, elements that Sperner's Lemma tells us must exist. By the same line of reasoning used in the one-dimensional case above, such an element must contain a fixed point of the transformation. So by doing the chopping fine enough, excellent approximations to the fixed points can be obtained. The interested reader can find more details on these procedures and why they work in the material by H. Scarf listed for this chapter in the bibliography.

## The Fixed-Point Property

Lots of important sets like closed line segments and closed squares and disks in the plane have the fixed-point property. That is, any continuous transformation of these sets to themselves has a fixed point. We already know that this fact is important in establishing the existence of things like equilibrium prices in an economy. But when faced with more complicated types of spaces, it's not always easy to prove directly that the space has the fixed-point property. So we're often forced into attacking the question indirectly by trying to prove that the original space of interest is topologically equivalent to one of the spaces like the square or the sphere, for which the fixed-point property has already been established. This is the way in which our earlier results on the topological equivalence of spaces make contact with fixed-point results like those obtained by Brouwer.

Suppose, for example, that we start with a continuous mapping $f$ of a closed ellipse in the plane to itself. We might recognize that this space is both compact and convex, thereby allowing us to apply the Brouwer Fixed-Point Theorem directly to establish the existence of a fixed point of $f$ acting on the ellipse. But in case we don't happen to notice these nice properties of the closed ellipse, there is another, less direct, way.

Algebraically, the ellipse consists of those points in the $x$-$y$ plane satisfying the inequality

$$\frac{x^2}{a^2} + \frac{y^2}{b^2} \leq 1,$$

where $a$ and $b$ are real numbers representing the semimajor and semiminor axes of the ellipse. (See Figure 2.12.) Now we deform this region by first letting $a$ smoothly go to $b$, and then letting $b$ smoothly go to 1 as shown in the middle of Figure 2.12. If we're careful in how we do these limiting operations, what we accomplish is to generate a sequence of new regions, each of which is topologically equivalent to the original ellipse. The end station on this sequence of deformations of the ellipse is just the unit disk

$$x^2 + y^2 \leq 1,$$

which we already know has the fixed-point property. But since this property is invariant under topological transformations, the original ellipse must then also have the fixed-point property. QED.

Ideally, however, what we'd like is to have some easy-to-test set of conditions that could be used to decide whether *any* given set does or does not possess the fixed-point property. Unfortunately, there appear to be no such criteria that are both necessary and sufficient. But that doesn't mean that we are completely powerless. For example, a good rule of thumb to employ is to demand that a set with the fixed-point property be both compact and contractible to a point. This latter condition means that we can continuously shrink the set to a point, just as a disk can be shrunk to a point by continuously reducing its radius to zero. Or think of a set of those Russian dolls called *matrioshka,* in which each doll contains a smaller copy of itself. If the outer doll is our space of interest, then shrinking

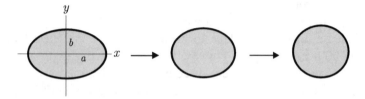

**Figure 2.12** Deformation of an ellipse into a circle.

76

this doll continuously would finally yield a doll that consisted of just one point. If such a shrinking is possible, then the original "mama" doll constitutes a compact, contractible set. If a set lacks either of the properties of compactness or contractibility, we can usually find a continuous mapping of the set that has no fixed point.

But not so fast! There exist sets that are neither compact nor contractible, yet *do* possess the fixed-point property. But these sets are pretty pathological. An example of this kind of set is shown below in Figure 2.13. Unfortunately, things can go in the other direction, as well. For instance, in 1953, S. Kinoshita produced the subset of $R^3$ shown in Figure 2.14 that is both compact and contractible—yet lacks the fixed-point property. So about the best we can say is that compactness and contractibility are topological indicators that a set probably has the fixed-point property. But the two conditions are neither necessary nor sufficient to guarantee it.

In view of our earlier deliberations, it's hard to argue that topological methods based on the notion of a fixed point are not crucially important in understanding what types of behaviors are and are not possible in

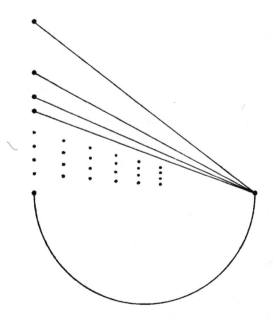

**Figure 2.13** A noncompact, noncontractible set with the fixed-point property.

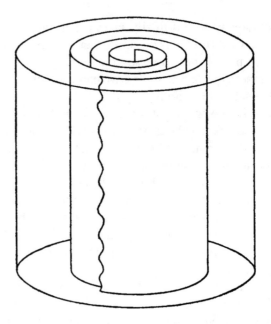

**Figure 2.14**   A compact, contractible set without the fixed-point property.

an economy. But it's not in economics alone that we find practitioners making use of fixed-point arguments to establish the existence of solutions to problems of concern. So to bolster our case for the importance of fixed points in human affairs, we conclude the chapter with few more examples of the many areas in which different flavors and dialects of Brouwer's pioneering result have been employed.

## Reaching for the Moon

On July 22, 1969, Neil Armstrong became the first human to set foot on a world beyond Earth. But behind all the media hoopla and NASA hype trumpeting the engineering and technological advances underpinning Armstrong's historic achievement, the real—but unsung—hero of the historic *Apollo 11* mission was none other than Newton's equations of motion for a particle moving in a gravitational field. Put crudely, the problem of transporting the *Apollo 11* space capsule from the Earth to the moon involved finding a solution of Newton's equations, subject to the condition that the trajectory should start at Cape Canaveral, Florida and end up on the moon on a plain near the southwestern edge of the Sea of Tranquility.

The moon-landing problem is what we term a *two-point boundary-value problem*. If we agree to let $q_1(t)$, $q_2(t)$, and $q_3(t)$ be the spatial coordinates of the *Apollo 11* capsule at any moment in time $t$, while labeling the velocity of the capsule in the three spatial directions by $p_1(t)$, $p_2(t)$, and $p_3(t)$, then the state of the capsule at any moment is characterized by giving just these six numbers. In other words, the time history of the *Apollo 11* mission is abstractly just a curve in the six-dimensional space $R^6$. The boundary conditions require that the spatial coordinates of the curve at time $t = 0$ coincide with those at Cape Canaveral, while at a fixed future time $t = T$ when the mission is to end, the spatial coordinates should be those of Tranquility Base. More-over, the velocity coordinates at the termination time should all be zero so that the astronauts don't have to make a crash landing. Finally, the initial velocity coordinates are just those given by the thrust of the Saturn rocket carrying the *Apollo 11* space capsule into orbit. So the mathematical problem becomes that of finding a curve in six-dimensional space satisfying the above conditions, and whose slope obeys Newton's second law (mass = force × acceleration) at all moments of time between $t = 0$ and $t = T$.

The first order of business is to convince ourselves that there ac-tually exists *some* curve satisfying the above desiderata. If more than one such curve exists, then we can argue about which path to the moon will minimize transit time, fuel expenditure, design costs, or any other criterion of goodness that the NASA engineers might dream up. Settling the existence problem is where fixed-point theorems come into play.

The collection of all smooth curves connecting two points in the space $R^6$ is an *infinite-dimensional* topological space $X$. In general, at most one such curve will have a slope that also satisfies Newton's equations of motion at each point in time. So we start by picking any curve from the set of all curves connecting the two boundary points. Call this curve $x_0(t)$. It will probably not satisfy our requirements, so we need to develop some procedure for improving upon the initial guess $x_0(t)$.

To do this, we first construct a mapping $\mathcal{F}$ that accepts a curve in the space of all smooth curves passing through the right endpoints and transforms it to another curve in this same set. Second, we want to have the curves that remain invariant under $\mathcal{F}$ (the fixed points) have positions and slopes that satisfy Newton's equations. Applying this transformation $\mathcal{F}$ repeatedly leads to the iterative process $x_1(t) = \mathcal{F}(x_0(t))$, $x_2(t) = \mathcal{F}(x_1(t))$, and so on. If the sequence of curves $\{x_0(t), x_1(t), x_2(t), \ldots\}$ converges to a definite curve $x^*(t)$, then that curve will be a fixed point of $\mathcal{F}$, hence, a solution to our problem.

From this sketchy outline, we see that we need two things in order to be able to appeal to the Brouwer Fixed-Point Theorem for ensuring the existence of a moon-landing trajectory: (a) an extension of the original result of Brouwer to the case of the infinite-dimensional topological space, whose elements are curves passing through Cape Canaveral and the Sea of Tranquility, and (b) a method for constructing the continuous mapping $\mathcal{F}$ of this space whose fixed point(s) are curves that are solutions of Newton's equations.

The first of these desiderata was satisfied through a major extension of Brouwer's result due to Pawel Schauder, who in 1930 proved an analogous fixed-point theorem for the case of infinite-dimensional topological spaces. The construction of a suitable mapping $\mathcal{F}$ involves expressing Newton's equations of motion in the form of an integral, rather than a differential, equation and then employing something called the Contraction Mapping Principle, which is a mathematical result very closely related to the Brouwer Fixed-Point Theorem. But to go into the details of exactly how this goes would require a bit more mathematical background on the part of the reader than we care to presume here. So let it suffice to say only that such procedures for constructing $\mathcal{F}$ do indeed exist, and are routinely employed, together with fixed-point arguments, in all areas of applied mathematics to establish the existence of solutions to a wide variety of important problems—and not all in the physical sciences. Here's another example of this sort, but this time from an area of the social sciences.

## Occupational Mobility

Sociologists are interested in the movement between different occupational classes from one generation to another. As a simple example, suppose there are three occupational classes: upper $(U)$, middle $(M)$, and lower $(L)$. Based on data gathered in England and Wales in the late 1940s, the graph in Figure 2.15 shows the probability of an individual

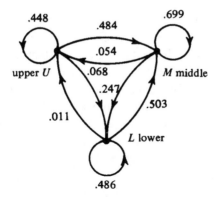

**Figure 2.15** Transition graph for intergenerational occupational mobility.

moving from one of these classes to another in a single generation. We can also represent these state transitions by the matrix

$$P = \begin{array}{c} \\ U \\ M \\ L \end{array} \begin{array}{ccc} U & M & L \\ \begin{pmatrix} 0.448 & 0.484 & 0.068 \\ 0.054 & 0.699 & 0.247 \\ 0.011 & 0.503 & 0.486 \end{pmatrix}, \end{array}$$

where we interpret the entry in row $i$ and column $j$ as the probability that a son or daughter of a person working in an occupation of class $i$ gets a job in an occupation of class $j$.

Suppose we let the vector $x(t) = (x_U(t), x_M(t), x_L(t))$ represent the fraction of the workers in each social class at time $t$. Then, by definition, $x_U$, $x_M$, and $x_L$ are all nonnegative numbers between 0 and 1, satisfying the constraint $x_U + x_M + x_L = 1$. The quantity $x(t)$ is what's called a *probability vector*. If $x(0)$ is the system's state at time $t = 0$, the rules of vector-matrix multiplication give the state at time $t = 1$ as $x(1) = x(0)P$. At time $t = 2$, the state is then $x(2) = x(1)P = x(0)P^2$ and, in general, the state at time $t$ is $x(t) = x(0)P^t$. Probably the single most important question we can ask is whether the state $x(t)$ converges to a definite distribution of workers $x^*$ after a sufficiently long period of time.

To answer this question, we observe that $x(t + 1) = x(t)P$. This implies that if the vector $x(t)$ approaches a vector $x^*$ as $t$ goes to infinity, then we must have $x(t)P$ approaching $x^*P$ since the matrix $P$ is independent of $t$. In other words, $x^*$, if it exists, must be a fixed point of the linear transformation represented by the matrix $P$, that is, $x^* = x^*P$. This transformation maps the space $S$ of all probability vectors to itself. But $P$ is a constant transformation of $S$ to $S$, hence continuous, and $S$ is a compact, convex subset of $R^n$. So the existence of such a point $x^*$ in $S$ follows immediately from the Brouwer Fixed-Point Theorem.

Let's finish off this example by suggesting that the reader verify that the limiting state $x^*$ in the occupational mobility example is $x^* = (0.067, 0.624, 0.309)$. We conclude from this that the most likely job for an individual to hold is a middle-class occupation, which is twice as likely an outcome as being in a lower-class job. And the likelihood is that only about 1 person in 16 will hold an upper-class position (naturally!). Now let's conclude our discussion of fixed points by showing how they enter into the debate with which we began this chapter of answering the question, "Who's Number One?"

## Top Dog

We opened this chapter by showing that the strength of a football team might plausibly be measured by a mathematical function involving non-negative quantities $a_{ij}$, which depended on the outcome of games played between teams $i$ and $j$. Forming these numbers into a matrix $A$, we further saw that producing a ranking vector $r$ of the teams came down to finding a solution to the equation $Ar = r$. Here we want to explore this matter further, the first item of business being to ensure that such a ranking even exists.

It turns out that under rather weak conditions on the entries of the matrix $A$, an important result in matrix theory, termed the *Perron–Frobenius Theorem,* tells us that this problem has a solution. Basically, the theorem states that there is a unique ranking vector $r$ having nonnegative entries that satisfies the equation $Ar = r$. The Perron–Frobenius result, in turn, follows from an application of the Brouwer Fixed-Point Theorem, which recognizes that a matrix like $A$, all of whose elements are nonnegative, is a continuous linear transformation of the set of nonnegative vectors into itself. Hence, there must exist at least one such vector $r$ satisfying the equation $Ar = r$. The Perron–Frobenius Theorem sharpens this result to tell us that under the conditions that apply in this type of problem, there is exactly one such vector.

Suppose we use the simple choice given earlier for the numbers $a_{ij}$, namely, that

$$a_{ij} = \begin{cases} 1, & \text{if team } i \text{ beat team } j, \\ \frac{1}{2}, & \text{if the game was a tie,} \\ 0, & \text{if team } i \text{ lost to team } j. \end{cases}$$

Now let's guess an initial ranking vector $r^{(0)}$, taking all the entries equal to 1. This means that at the outset we assume all teams are of equal strength. Then the $i$th component of the vector $Ar^{(0)}$ becomes simply the winning percentage for team $i$. By the same token, the $i$th component of the vector $A^2 r^{(0)}$ is the average winning percentage of the teams defeated by team $i$. This vector presumably contains valuable information about the strength of team $i$'s schedule. Is this a good measure to use for determining a national champion? Well, it's better than the simple winning percentage, the vector $Ar^{(0)}$. But it's still not good enough.

The choice given above for the entries $a_{ij}$ leaves a lot to be desired, especially in sports like football where teams play each other only once, or occasionally twice (in a bowl game), during a given season. For example, whether the game is very lopsided or nearly even, this assignment of the numbers $a_{ij}$ gives all the credit to the winner. Furthermore, beating a winless team is more damaging to your rating than not playing that team at all, since the winning team earns no points for beating a winless club and increases its number of games by 1. This comes about because the ranking scheme assigns a rank of zero to teams that are winless.

A better way to assign the quantities $a_{ij}$ would be to distribute the 1 point between the competing teams in a continuous fashion. For example, if team $i$ scored $S_{ij}$ points in a game against team $j$, while team $j$ scored $S_{ji}$ points in the same game, we might assign the value $a_{ij} = (S_{ij}+1)/(S_{ij}+2)$ to team $i$. But this scheme has its own weakness, namely, that to get a high ranking a team can show no mercy. It should run up as big a score as possible, even when the game is already out of hand. So to avoid having teams "pile it on" to improve their ranking, we should distribute the 1 point of credit in a nonlinear fashion. Good ways to do this are discussed in the material cited for this chapter in the bibliography. However, even these assignments have the weakness that they place too much emphasis on the strength of a team's schedule. This is because a team can never earn enough points playing weaker opponents to increase its score so as to get a high ranking. So what to do?

One way of addressing the "strength-of-scheduling dilemma" was proposed recently by mathematician James Keener. His idea is to calculate the rank for each team as a *nonlinear* function of the outcome of each game. More specifically, instead of the linear rule (see Eq. (∗) in the section "Needles and Haystacks" above), with its *constant* quantities $a_{ij}$, Keener used a formula in which these quantities themselves could depend on a team's ranking. This led to a *nonlinear* relationship between the this week's ranking and the next.

After all the mathematical smoke clears away, to find a ranking vector $r$ by Keener's scheme it's necessary to solve a nonlinear equation $r = F(r)$, where $F$ is a continuous—but nonlinear—map of the earlier hypersquare to itself (this time with sides of length 1 instead of 100). But we know that the Brouwer Fixed-Point Theorem holds for a square, so we can use it assert that there must exist at least one vector $r$ that is a fixed point of the transformation $F$. So Keener's problem has a solution. Using this nonlinear scheme, Keener obtained the rankings shown in the next-to-last column of Figure 2.1 for the 1993 college football season.

To conclude, it's interesting to compare this mathematical procedure with the rankings of the sportswriters and the coaches. First of all, there is much more variation between the rankings from the Keener system and that of the media polls than between the rankings of the polls themselves. This suggests that the media polls are far from independent. The results also indicate that there is really no "best" type of ranking system; different schemes give rise to different results because they weigh key factors differently. And when you "tweak" one scheme to get rid of some bad feature, you invariably end up introducing a new counterintuitive aspect that you'd like to get rid of. Finally, after several years of studying these methods, Keener feels that intuition is a pretty poor guide to determining a ranking. It seems that with over a 100 major college football teams in competition, there are just too many factors for intuition to be able to adequately account for all of them.

# CHAPTER

# 3

# Morse's Theorem

*Singularity Theory*

# That's the Way the Paper Crumples

Japanese culture is known for its many refined and sophisticated forms—haiku poetry, Zen Buddhism, sumo wrestling, bonsai trees, kabuki theater—differing widely from their Western counterparts. Another item on this list is the delicate art of paper folding, or what in Japanese is termed *origami*. This involves folding a piece of paper in such a way that when unfolded the paper springs forth into a completely unexpected shape, usually some sort of animal. For instance, Figure 3.1 shows a flying bird, together with the flattened square of paper from which it emerged. Mathematically speaking, origami involves the transformation of a two-dimensional object into a three-dimensional one, the fold lines on the paper being the two-dimensional projection of the emerged, folded object onto a flat plane. The basic question for origami aficionados is: Given a particular three-dimensional "target" figure, what are its projections onto a square of paper? In other words, where and how does one have to fold the paper in order to produce a particular object? As such, origami belongs to the field called *projective geometry*.

Our concerns in this chapter are with describing the geometrical shapes—abstract and concrete—that we can expect things like a living organism or processes like the movement of stock prices to take when they proceed in accordance with the laws of physics, biology, engineering, economics, or whatever. Such real-world phenomena are characterized by quantities like the concentration of various chemicals in a cell or economic factors like interest rates and corporate profits in a financial market. The actual physical shape of a collection of cells (that is, an organism) or the abstract shape of a security price history is then determined by the relationship(s) among these factors. Speaking slightly more technically, the shapes are a *function* of these variables.

In the majority of cases, if we change the values of the chemical concentrations or the levels of the economic factors just a little bit, the "shape" of the system also shifts only slightly. But sometimes there is a particular configuration of values that if we move away from them slightly, the resultant shape of the organism or stock price chart can change dramatically from, say, a male to a female, or from a period of steadily rising prices to a market crash. To illustrate, an economy might have several states that it can be in—boom, recession, slow growth, stagflation, or depression. Suppose these states are the end result of the

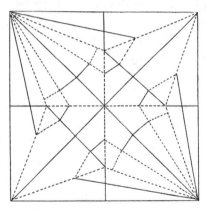

**Figure 3.1** Origami of a flying bird.

values of variables like money supply, unemployment, foreign balance of payments, and consumer confidence. Thus, the particular state the economy is in is a function of the values of these quantities. The values of these variables at which a small change away from these values can move the economy from one of these qualitatively different states to another is then a critical point of the function characterizing the economy. It is a place where a small change in the "inputs" to the economy may give rise to a dramatic shift in its "output."

An important part of our story in this chapter is about how to mathematically pin down where the critical regions of such chemical concentrations or economic factors are, and just what *kinds* of discontinuous shifts are likely to arise if we move into these "danger zones." Since the mathematical descriptions of physical and social phenomena are by functions, our job here is to try to characterize the behavior and the geometrical forms of "typical" functions.

To see what this means, think of a whirlpool—perhaps the water running down the drain of a bathtub. Now suppose there's a bit of soap floating in the water near the drain. Once the soap comes near enough to the whirlpool, the geometry of its trajectory—direction of movement, velocity, and the orientation of the spiral—will be governed by the whirlpool's motion, even if the soap is not actually in the whirlpool itself but only nearby. The point at the center of the whirlpool is an example of what we mean by a *critical point,* in this case a critical point of the function describing all the fluid flow in that region of the tub. The critical point in this bathtub example is the point to which all nearby points are attracted or "sucked in." But in a different context the critical point might be a point from which all nearby points are repelled. Or it could even be a point at which the system undergoes an even more complicated type of local behavior. We'll see the possibilities as we move through the chapter.

From this whirlpool example, it's easy to see that it's the nature of the critical point—its strength, scope, and direction of rotation—that dictates the paths of objects floating nearby, that is, the local behavior. It turns out to be the same in mathematics, too. The behavior of a function at one of its critical points tells the story of how the function behaves at all nearby points. It is one of the principal goals of this chapter to establish exactly what this means, mathematically speaking, and to pin down precisely the qualitatively different types of local behaviors that can arise near a critical point.

A key element in identifying the local behavior of a function near a critical point is to strip away all the aspects of its behavior that might be attributable to accidents in the way we happen to choose to write down the function. So in the stock market situation, whatever the function is that generates prices from variables such as interest rates, corporate earnings, and the like, the intrinsic behavior of that function should certainly not depend on whether we happen to measure prices in dollars or yen or whether the current interest rate is expressed in percentages or in points. In short, what we're looking for is the properties of the function that "look the same" in *any* coordinate system we happen to use to explicitly describe the function.

The reader will recognize this problem as the same one we faced in Chapter 2, where what was required was to find the properties of a closed surface that remained unchanged when we squeezed, stretched, twisted, or deformed the surface in any other way that didn't tear it or punch a hole in it. The situation is the same here, except that now it is abstract spaces whose "points" are functions that will be transformed to themselves instead of spaces of geometrical points. What we'll be trying to identify is the "simplest" function that displays the behavior characteristic of a whole class of functions, just as the sphere with, say, two handles served to represent completely the topological structure of *all* orientable, closed surfaces of genus 2.

Historically, the line of research that we focus upon in this chapter arose with a question that also relates to paper folding. Suppose I take a piece of a rather flexible type of paper—tissue paper, for example—and crumple it up however I please, the sole restriction being that no creases are allowed. Figure 3.2 shows what a small section of this crumpled ball would look like if we were to examine it under a magnifying glass. Suppose now that a small insect (a fly, perhaps) walks around on the surface of the wadded-up ball of paper. As the fly makes its way along the surface, it encounters points like *a*, near which the surface is flat, those like *b* that lie on the edge of a "cliff" and yet other types of points like *c* and *d*, which are geometrically even more complicated. So one might well ask: If there's nothing special about the way the paper has been crumpled, that is, no constraints have been imposed limiting the way it can be wadded up other than that it have no creases, what types of points can the fly *expect* to encounter as it wanders about on the wadded-up ball of paper? In other words, what are the "typical" ways to scrunch up a piece of tissue paper?

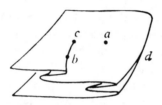

**Figure 3.2**   Local region on a crumpled ball of paper.

To understand this question more fully, let's be a bit more specific about what we mean by a "crease." The exclusion of creases means simply that the twists and turns in the paper are what mathematicians call "smooth"; so unlike in origami, where sharp creases are crucial to the unfolding of many forms, here we forbid putting knife-edged creases in the paper. Under this condition, there are, generally speaking, three possible shapes that the crumpled ball can assume near any point:

a. The paper lies flat near the point.
b. The point lies on a fold line of the paper.
c. A pleat is being formed at the point.

The points labeled $a$, $b$, and $c$ in Figure 3.2 illustrate each of these types of points, respectively.

It doesn't take too much experimentation with wadded-up balls of paper to discover that other types of points can also turn up. For instance, the point $d$ in Figure 3.2 is a new type. But with a little more experimentation in twisting and folding the paper, we quickly discover that an arbitrarily small change in the position of the paper can make all points disappear if they are not of the types $a$, $b$, or $c$. For example, Figure 3.3 shows how the point of type $d$ disappears as soon as the bottom fold is pulled just a little bit to the right. On the other hand, points of types $a$, $b$, and $c$ cannot be made to disappear by small changes of this sort. So these are the types of points that a fly would expect to encounter as it wanders around on the surface of a crumpled piece of paper. In more formal terms, crumplings of the paper that contain only these kinds of typical points are termed *generic*, since any other crumpling can be transformed into a generic one by an arbitrarily small distortion of the original ball of wadded-up paper.

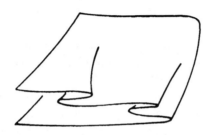

**Figure 3.3** A generic folding of the paper.

91

Not only can we see that points of type $a$, $b$, and $c$ are the typical possibilities, we can also see what makes them typical: they are *stable*, in the sense that a small perturbation of the wadded-up ball of paper cannot make them disappear. Points like $a$ are the simplest and most common, so we term them *regular* points. All other points are called *singular*, because they form a very small subset of the set of all possible points. Among the singular points, only those like $b$ and $c$ are stable. Thus, our hypothetical fly roaming about on the surface of the tissue paper can expect to find regular points, fold points, and points at which two fold lines come together (technically, these are called *cusp* points). If it encounters any other type of point, the fly has a right to feel surprised. Moreover, that surprising type of point can always be transformed into a point like $a$, $b$, or $c$ just by jiggling the paper a little bit.

The process of smoothly wadding up a piece of tissue paper is an example of a transformation of one plane, the original flat piece of paper to another, the wadded-up ball. (Recall: From Chapter 2 we know that a flat piece of tissue paper is topologically equivalent to a wadded-up piece of paper—provided there are no tears in the paper. Thus, the wadding-up process is mathematically the same thing as transforming a plane to itself.) So this is an example of a function that assigns one or more new points in the plane to every old one. And in this sense, just like with the economic examples above, one can regard the original flat sheet of tissue paper as the "input" to the transformation, the "output" then being the wadded-up ball that's probably in your wastebasket by now.

The arguments given above constitute an informal statement of Whitney's Theorem, which was proved by Hassler V. Whitney in 1955. What he showed was that for any smooth transformation (that is, one with no creases) taking the points of a plane into another plane, the only kinds of points that can typically turn up are regular points, folds, and cusps; any other type can be made to disappear by just changing the transformation a little bit.

While this result seems rather special, pertaining as it does only to smooth transformations of planes into planes, it opens up a plethora of questions since if one can prove that there are only three different types of points that can typically occur for maps of planes to planes, then perhaps one might also be able to prove analogous results about smooth transformations of other kinds of spaces to each other. For example, smooth transformations of the everyday three-dimensional world of $R^3$

to the real line $R^1$, or transformations of the surface of a doughnut, the torus $T^2$, to another torus. Why anyone would want to know about such things is also an important part of our story here, and will be answered as we go along. If we had this kind of information, then we would be able to display the standard, that is, simplest, way these transformations can arise, as well as gain insight into how the real-world phenomena they purport to represent might behave as we shift the variables in the problem by small amounts.

In fact, Whitney's result did not come out of the blue. It was motivated by earlier investigations by Marston Morse in the 1930s on the nature of the critical points of smooth *functions,* rather than mappings of the plane. So what Morse considered was smooth transformations of ordinary $n$-dimensional space into the real numbers, a much smaller space than a plane. Morse was able to show that near a critical point of such a function, we can expect the function to look geometrically like a saddle—curving upward in some directions and downward in the others. Moreover, he showed that this geometry is preserved under any small changes in the function, that is, such functions are *stable* in the sense that all nearby functions give this same geometrical picture near the critical point in question. But before describing these results in greater detail, let's have a look at a particular problem from physics illustrating what Morse was able to establish in much greater generality.

## Fluid Flow Between Two Cylinders

Consider the case of what's called a *two-roll mill,* shown in Figure 3.4. This consists of two rollers placed in a fluid and set rotating in the same direction. Versions of this problem arise regularly in mechanical engineering whenever there are shafts rotating in a fluid. For example, the camshafts of most car engines rotate in this fashion, with the fluid being the car's motor oil. It's important to be able to calculate the stream lines in the fluid, given the rate of rotation of the rollers and properties of the fluid like its viscosity, since it's when the stream lines break down that turbulence sets in. And the onset of turbulence is exactly when the lubricating effect of motor oil grinds to a halt (literally!).

From Figure 3.4, we see that there is a stagnation point between the rollers at which the flow is zero. This comes about because the fluid moves in opposite directions but with the same velocity between

93

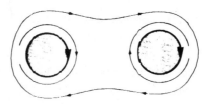

**Figure 3.4**  A two-roll mill.

the rollers. Thus, the motion of the fluid moving up in the middle is canceled by the motion of the downward-moving fluid, leading to a point of zero flow—the stagnation point. It turns out that the stream lines of a flow pattern near the stagnation point in the $x$-$y$ plane can be mathematically characterized by taking different values of $c$ in the equation $ax^2 + by^2 = c$, first fixing the numbers $a$ and $b$ to have opposite signs. Such a set of stream lines is shown in Figure 3.5. We note that the real-world stream lines into and out of the stagnation point are actually curved. Therefore, to reduce the true stream lines of the flow to the idealized model in Figure 3.5 would require a change of the variables $x$ and $y$ so as to "straighten" the stream lines into those passing through the stagnation point at the center of the figure. Part of Morse's Theorem is devoted to assuring us that such a "straightening" change of variables always exists.

But even if we remain in the original $x$ and $y$ coordinates, the simple approximation $ax^2 + by^2$ works very well—even for physical

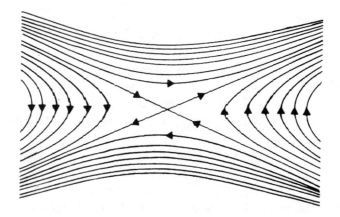

**Figure 3.5**  Theoretical stream lines for the two-roll mill.

calculations involving the flow near the stagnation point. The success of this approximation is mostly attributable to the fact that the standard, or typical, way for stream lines to look is given by this quadratic function. Physically, this is because the energy in the fluid is minimal with this configuration, a fact that follows from the principle that every system moves to its lowest energy state unless there is some other factor preventing it from doing so. Mathematically, the reason this quadratic is the simplest model is a direct consequence of Morse's Theorem, which states that near a critical point of a function (here the stagnation point of the function describing the stream lines of the flow), it is typical for a function to look like a quadratic. So, if there are no special constraints imposed on the flow, the stream lines can be expected to look like those shown in Figure 3.5. This means that a hydraulic engineer trying to mathematically picture the flow near the stagnation point doesn't need to search through all possible functions to create a model of the flow. The simplest model is already at hand—and it's pretty simple, just a quadratic. And how does the engineer know this is the simplest model? Easy—Morse's Theorem!

To explain why this is the case, as well as to set the stage for an exposition of Morse's work, we have to speak briefly about how to mathematically represent smooth functions like the one above describing the stream lines in the two-roll mill. The values of every smooth function near a given point can be expressed as an infinite series in powers of the function variables. For example, the familiar sine function from elementary trigonometry is given near the point $x = 0$ by the series

$$\sin x = \frac{x}{1!} - \frac{x^3}{3!} + \frac{x^5}{5!} - \frac{x^7}{7!} + \cdots,$$

where ! denotes the factorial function $x! = x(x - 1)(x - 2) \cdots (2)(1)$. This series allows us to compute good approximations to the values of $\sin x$ for values of $x$ near $x = 0$. By looking at the terms in such a series, we can identify just how far out in the series we have to go in order to understand *everything* about how the function looks near the origin. As we'll see, Morse's Theorem says you usually don't have to go very far; in fact, you don't have to go beyond the second-order terms unless there's something very special about the function. In those cases when there *is* something special about the function, catastrophe theory tells us how much farther we have to go. The idea that both these theories rests upon is showing how to eliminate *entirely* the terms represented by "···" through an adroitly chosen coordinate transformation. So let's have a look at the arguments showing how this can be done.

## A Taylor's Tale

Suppose we have a function $f$ of a single variable $x$, which we can represent as the smooth curve shown in Figure 3.6. If the curve is smooth enough (which will be our standard operating assumption throughout the remainder of the chapter), then it is a known fact from elementary calculus that for any point $x$ near a given base point $x_0$, the value of $f$ at $x$ can be expressed by the infinite series

$$f(x) = \alpha_0(x_0) + \alpha_1(x_0)(x - x_0) + \alpha_2(x_0)(x - x_0)^2 + \cdots, \quad (*)$$

where the quantities $\alpha_i(x_0)$ are real numbers whose values depend on the chosen base point $x_0$ (and, of course, on the function $f$). This infinite-series representation is called the *Taylor series* for $f$ at the point $x_0$, while the numbers $\alpha_i(x_0)$ are called the *expansion coefficients*. (Technical aside: For those familiar with calculus terminology, these expansion coefficients are, up to a constant multiplier, simply the derivatives of the function at $x_0$. But we will not need to make use of this fact here.)

The first thing to note about the Taylor series $(*)$ is that it is a *local* representation; it gives the value of $f$ at points $x$ that are near the base point $x_0$. If we want to know what the function is doing at a point $x$ far

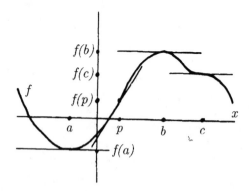

**Figure 3.6**  A smooth function of a single variable.

away from $x_0$, then a new infinite series must be computed at a new base point near that distant point. And this new series will, of course, have its own set of expansion coefficients.

It simplifies our life—as well as our notation—to assume henceforth that our concerns are with the behavior of $f$ near the origin, that is, we take the base point to be $x_0 = 0$. This involves no loss of generality, since we can see from Figure 3.6 that by sliding the $x$-axis horizontally it's always possible to arrange to have the base point coincide with the origin of the coordinate system. Similarly, by sliding the vertical axis representing the values of $f$ up or down, we can also arrange for the curve $f$ to pass through the origin. This means that we can take $f(0) = 0$. Putting these two preliminary coordinate changes together—from $x_0$ to 0 and from $f(x_0)$ to 0—the Taylor series for $f$ at the origin can be written as

$$f(x) = \alpha_1(0)x + \alpha_2(0)x^2 + \alpha_3(0)x^3 + \cdots , \qquad (**)$$

which is valid for all points $x$ near the origin. And to further streamline our notation, let's drop the argument 0 from the expansion coefficients, now that we have agreed that our concerns will always be with the local behavior of $f$ near the origin.

Suppose now that $\alpha_1$ is not equal to zero. Then the expression $(**)$ suggests that $f(x)$ will be approximately equal to $\alpha_1 x$, since the higher-order terms can all be made as small as we like simply by taking $x$ near enough to 0. This is because if $x$ is very small, then $x^2$ will be much smaller than $x$ and $x^3$ will be smaller still, and so on for the higher powers of $x$. Consequently, if we choose a point $x$ that is *very* close to 0, the term $\alpha_1 x$ will dominate the rest of the terms in the infinite series. In this situation, we call the origin a *regular point* of $f$, for much the same reasons that the point $a$ of Figure 3.2 in the paper-folding problem was called a regular point, namely, that these are the typical kinds of points that a smooth function $f$ can have. The point $p$ in Figure 3.6 is an example of such a regular point.

But what if $\alpha_1 = 0$? Point $b$ in Figure 3.6 serves as an example of this kind of critical point, near which the curve representing $f$ is "bowl-shaped." The direction in which the bowl points—up or down—is determined by the sign of $\alpha_2$: positive means the bowl points up, negative means that it points down. Recalling the discussion about fluid flow, if $\alpha_2 > 0$ such a point represents the fluid configuration when it is in its minimal-energy "shape." Moving away from this shape would then be like trying to climb up the sides of the bowl. In such cases, we call the origin a *nondegenerate critical point* of $f$. So when $\alpha_1 = 0$, the expansion (∗∗) suggests that for $x$ sufficiently close to 0, we have $f(x) \approx \alpha_2 x^2$. In other words, near a nondegenerate critical point the function $f$ is approximately equal to its quadratic part.

But it could happen that $\alpha_1$ and $\alpha_2$ both equal 0. In that event, we call the origin a *degenerate critical point*. The point $c$ in Figure 3.6 is an illustration of this sort of point. One part of Morse's Theorem tells us that this case is unusual; nevertheless, it does happen. And when it does, it generally means that strange things are going on with the physical process that the function represents, things like the buckling of a beam or the outbreak of a political revolution. When both $\alpha_1$ and $\alpha_2$ are zero, the expansion (∗∗) says that if $x$ is small enough we should have $f(x) \approx \alpha_3 x^3$. And so it goes, the local behavior of $f$ near the origin being approximated by the first nonzero term in its Taylor series representation.

For smooth functions, the key property of interest is how they behave near a critical point since near a regular point the behavior is, well, boringly regular—essentially just that of a straight line (or a flat plane in higher dimensions). But things start to happen at a critical point, since that's where the function "turns," moving from being a local minimum or maximum, or perhaps doing something even more strange if it's a degenerate critical point. So if we want insight into how the function is put together from its various local pieces, we need to uncover information about its behavior near the critical points. Any critical point of $f$—degenerate or nondegenerate—is called a *singularity* of the function. It is the goal of singularity theory to make mathematically airtight the foregoing plausibility arguments about the local behavior of a function, as well as extend them to cases when $f$ is a function of $n$ variables, $x_1, x_2, \ldots, x_n$, instead of just the single variable $x$ we have been considering thus far.

To see geometrically why the local behavior is so important in piecing together the overall global structure of a function, suppose we know that the critical points of a smooth function $f$ are located at the points $x = a$, $x = b$, and $x = c$. Furthermore, let's assume that $a$ and $b$ are nondegenerate critical points, such that $f(x) \approx x^2$ near $a$ and $f(x) \approx -x^2$ near $b$. Finally, let $c$ be a degenerate critical point, such

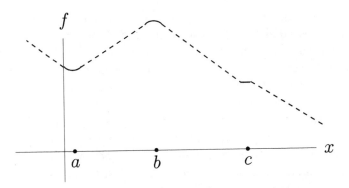

**Figure 3.7** The behavior of a function near three critical points.

that $f(x) \approx x^3$ in the vicinity of $c$. Graphically, this implies the situation shown in Figure 3.7. But since we know that the function's behavior is like a straight line away from each of the three critical points, we can use this fact to sketch in the dotted part of the curve. Thus, all we need to know in order to recreate the entire *global* behavior of the function is information about how it behaves *locally* near its critical points. Singularity theory gives us this information. To show how singularity theory works in helping us identify the places in a system where something funny could happen, let's look at the dynamics of an electric power generation system.

## Electrical Power Generation

Consider an electrical power supply network with two generators. A description of the power supplied by the generators in terms of the angular speed of the rotors in the generators and their torque angles is given by the function

$$V(x_1, x_2) = -c + \frac{1}{2}ax_1^2 - bx_2 - c \cos x_2,$$

where $x_1$ is the difference in the angular speed between rotor 1 and rotor 2, $x_2$ is the difference in the electrical torque angle of the two rotors, $a$ is the product of the angular momentum of the two rotors, $b$ is the difference in the damping factor between the rotors, and $c$ is the product of the sum

99

of the two angular momenta and the product of the voltages of the two generators. Here we identify the critical point of $V$ at the origin with the equilibrium position of the dynamical process for the motion of the two rotors. We shall say more about this kind of identification later, as it forms the point of contact between catastrophe theory and the bifurcation theory of dynamical systems. In this power generation situation, we'll see that singularity theory gives us a look at those combination of values of $a$, $b$, and $c$ that can lead to an abrupt breakdown in the supply of electricity to a region, perhaps even something as dramatic as the summer 1977 blackout that shook up New York City for several days. Thus, we would like to know about the behavior of the function $V$ when the rotors of the generators are near their equilibrium position at 0.

To analyze the nature of the critical point of $V$ at the origin, we write the function as a Taylor series. Unfortunately, to do this involves knowing a bit of calculus. So you'll have to take it on faith that the Taylor series has the following form for $x_1$ and $x_2$ near the origin:

$$V(x_1, x_2) = (a \cdot 0)x_1 + (-b + c \sin 0)x_2 + ax_1^2 + (0)x_1 x_2$$
$$+ (c \cos 0)x_2^2 + \cdots,$$
$$= -bx_2 + ax_1^2 + cx_2^2 + \cdots.$$

Therefore, the origin is a regular point if $b \neq 0$, a nondegenerate critical point if $b = 0$ and a degenerate critical point if $a$, $b$, and $c$ are all zero. In this latter case, of course, the function $V$ itself degenerates into the zero function.

In the 1930s, Marston Morse proved a pivotal theorem enabling one to generalize the fairly straightforward result given earlier to the effect that the lowest-order nonvanishing term in the Taylor series tells about the local behavior of a smooth function to functions of many variables. Here, for simplicity, our attention will be confined to the case of a finite number of variables, although under suitable conditions Morse's result can even be extended to infinitely many. We have already seen an example of such a function of two variables in the two-roll mill problem, where the energy in the system is given by the function $f(x, y)$ of the two spatial locations $x$ and $y$. The economic illustration briefly discussed above also leads to functions of several variables, as do problems of cellular differentiation and morphogenesis that will be discussed in more detail below in the section "The Shape of Things."

To describe Morse's Theorem calls for us to be a bit more precise about what we mean when we say that one function "looks like" another. This, in turn, entails introducing a notion of *smooth equivalence* into the space of smooth functions, in complete analogy to the way the concept of topological equivalence was introduced into the space of closed surfaces in Chapter 2.

## Tugging on Taylor's "Tayl"

Consider again the case when $f$ is a smooth function of a single variable $x$, and the origin is a nondegenerate critical point of $f$. Then we can write

$$f(x) = \alpha_2 x^2 + \text{Tayl},$$

where "Tayl" represents all the higher-order terms in the Taylor series expansion for $f$ near the origin, the so-called "tail" of the Taylor series. By neglecting Tayl, that is, by taking $x$ to be *very* close to the origin, we can often get a good approximation to the values of $f$, as already discussed. But wouldn't it be nice if we could make Tayl disappear *completely*? Then we would have an *exact* local representation for $f$ instead of just an approximation. Singularity theory shows us how to get rid of Tayl. And this is important because it may be the case that the local character of $f$ will not be captured by any *finite* number of terms in its Taylor series.

For example, consider an analytic function $f(x_1, x_2)$ of two variables, whose Taylor expansion up through terms of order 17 is

$$f^{[17]} = x_1^2 x_2 + \frac{1}{2} x_1 x_2^2 + \cdots + \frac{1}{15!} x_1^2 x_2^{15}.$$

Thus, any point on the $x_1$ or the $x_2$ axis is a solution of the equation $f^{[17]} = 0$. Now suppose we take one more term in the Taylor series of the function $f$, finding that the terms through order 18 are

$$f^{[18]} = f^{[17]} + \frac{1}{18!} x_2^{18}.$$

The equation $f^{[18]} = 0$ has no solution with $x_2 > 0$. This means that $f^{[17]}$ is not sufficient to determine even the character of $f^{[18]}$, let alone the character of $f$ itself. By showing us at exactly what point in the Taylor series we can make Tayl disappear, singularity theory tells us precisely how much of the Taylor series we need to know in order to completely describe the local behavior of a function.

Basically, the idea is to recognize that the choice of coordinate system is arbitrary. This is a bit like measuring people with meter sticks or yardsticks. You get a different number when you use different scales. But the height of the person is completely unchanged by this arbitrary choice of the measuring sticks. And so it is with functions, too. The essential character—the structure—of a function, what sets it apart from other functions, doesn't (or at least shouldn't) depend on the way we happen to describe the function through our choice of coordinates. So perhaps if we move from the original $x$ coordinates to an adroitly chosen new set of coordinates $y$, the function $f$ expressed in the $y$ coordinates will have Tayl $= 0$. In that case, we can feel confident that what remains is the essential structure of the function and not something that is a mere artifact of how we happened to choose the coordinates. Showing how to make Tayl disappear and seeing what's left is a large part of what singularity theory is all about. Now let's see what all this talk about coordinate changes has to do with two functions being equivalent.

## Look-Alikes

Finding the right (that is, useful) notion of equivalence for a set of mathematical objects depends entirely on how much distortion of the objects we're willing to tolerate and still regard them as being "the same." For the closed surfaces and other topological spaces considered in Chapter 2, where our concerns centered on the properties of objects that remain unchanged when we can do anything we like them except tear or cut them, it turns out that the right notion of equivalence is continuity: two objects or spaces are equivalent if they can be continuously deformed one to the other. But the situation changes when the objects are smooth functions.

Certainly one of the most general ways we can perturb, or deform, a smooth function is by adding a small continuous function to it. But deforming a function in such a way goes a bit too far in the direction of generality, since the critical points can be (and usually are) transformed to points that are no longer critical points of the new function. So we need a stronger notion of equivalence, one that leaves the nature (that is, degree of degeneracy) of a function's critical points unchanged. It turns out that the kind of equivalence that works best here is what's called *smooth equivalence:* two functions are smoothly equivalent if there is a smooth coordinate transformation (technically, a *diffeomorphism*) that converts one function into the other.

At this point it's worth reemphasizing that our concerns are with the *local* character of the function near a particular critical point, not its overall global behavior. This means that our coordinate transformation need only work in a small neighborhood surrounding the critical point. To demonstrate the difference, Figure 3.8 shows an example involving the two functions $f(x) = x^2$ and $g(x) = x^2 - x^4$. If one changes from the $x$ variable to a new variable $y$, defined by the smooth transformation

$$y(x) = \frac{x}{|x|} \sqrt{\frac{1 - \sqrt{1 - 4x^2}}{2}},$$

then $g$ is transformed to $f$ since $g(y(x)) = f(x)$ for all values of $x$ in a local region of the critical point at $x = 0$. However, no smooth change of coordinates can *globally* transform $g$, with its three critical points, to a function $f$ that has only a single critical point.

Morse used the idea of smooth equivalence to find the simplest types of smooth functions, to pin down just how many types there are, and then to show that a small deformation of any such function does not

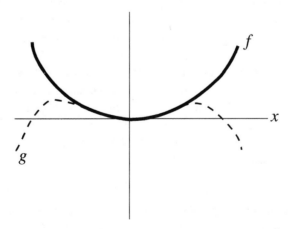

**Figure 3.8** Local equivalence of two functions.

change its type. In accomplishing all this, Morse's results allow us to characterize the behavior of a very large class of smooth functions. How large? Well, large enough to densely fill the set of *all* smooth functions. Morse's Theorem gives the whole story.

## Morse's Theorem

The question answered by Morse's Theorem can be thought of in the following way. Suppose we take the set of all possible smooth functions, put them in a bag, and shake it up. We then close our eyes and pull some function out of the bag at random. What kind of properties can we expect this function to have? In other words, what does a typical smooth function look like? For the answer, let's "deconstruct" the question into its component parts.

First, a typical smooth function will certainly have some critical points. Why? Well, because the only functions that do not are functions like $\tanh x$ that don't "wiggle" enough to ever make their graphs horizontal, or functions with no wiggles at all like straight lines and the exponential function $\exp x$. But such functions form an infinitesimally small part of the set of all possible smooth functions. Thus, we can expect a typical function to have some critical points. But what kind?

If we take out a magnifying glass and look at the structure of a typical function near one of its critical points, we can expect to find that the critical point is nondegenerate. This means that the image we see in the magnifying glass will not be a "flat" curve, but will look like a bowl, or in more than one dimension, a saddle. In short, the surface representing the function values is locally curved. One way of looking at this situation is to note that a very small change in a function $f$ near a degenerate critical point produces either a new function $f_1$, in which the degenerate critical point has been replaced by two nondegenerate ones, or a function $f_2$ having no critical points at all. This situation is shown in Figure 3.9 for the function $f(x) = x^3$ near the critical point at the origin.

Such a situation might occur in the real world if, for example, $f(x)$ represents the rate of production of, say, a hormone in your body. Suppose the production is locally "flat" when the body temperature $x$ is at or very near its normal level, which we can take to be $x = 0$ by

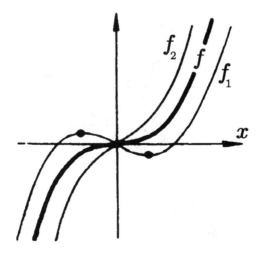

**Figure 3.9**  Dissolution of a degenerate critical point.

choosing a suitable scale by which to measure it. But the actual curve of hormone production might not be the theoretical one given by $f$ in Figure 3.9, but a nearby curve like $f_1$. In this case, hormone production would peak at a temperature slightly less than $x = 0$ and drop to a minimum when the temperature rose to slightly above 0. On the other hand, if the real curve describing hormone production turned out to be $f_2$ instead, then there would be no maxima or minima at all near 0, but just a steady increase in production as the temperature increased. The information telling us which of these is the case might then be used to determine the dosage and timing of a drug that regulates the production of the hormone.

If $f$ is a function of more than one variable its local geometry near a nondegenerate critical point looks like a saddle, since the graph of the function is a surface that may bend in different directions at a given point. Since the critical point is nondegenerate, this saddle must then curve downward in $k$ coordinate directions and upward in the remaining directions. The integer $k$, the number of downward curving directions, is called the *index* of the critical point. Figure 3.10 shows a function $f(x, y)$ of two variables near a critical point, where we can see it bending upward in the $y$-direction and downward in the $x$-direction. Thus, the function has index $k = 1$ at this critical point.

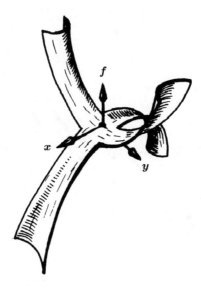

**Figure 3.10**   A function of two variables having index 1.

Finally, we consider the values a typical function assumes at its critical points, what are called its *critical values*. Typically, we can expect each of these values to be different. The reason is that if we have a function with two equal critical values, an arbitrarily small perturbation of the function will produce a new function whose critical values are all different. The reader can easily see this fact by thinking of the function values as representing the height of a range of mountains. If one looks at the range in profile, the slopes and peaks of the mountains form a graph of a function. If two peaks have *exactly* the same height, it would take only a very small amount of work with an ice ax on one of them to make the two heights differ by a small amount. It's equally easy to see that jiggling a function having distinct critical values results in a new function that also has distinct critical values. So the property of having distinct critical values is preserved under small changes of the function.

These arguments make it plausible to claim that a small, smooth perturbation of a function with nondegenerate critical points will yield a function having these same properties. It's equally clear that a similarly small, smooth perturbation of a "bad" function, that is, a mathematically more complicated function having either a degenerate critical point and/or equal critical values, produces a function without these analytic blemishes. The function $f(x) = x^3$ depicted above in Figure 3.9 shows why. Think of the graph of the function as a rubber band. By pushing down in one place and pulling up in another, a degenerate critical point like the one at the origin, where the function is locally flat, can be given small "humps," thereby dissolving the critical point into two nearby non-degenerate critical points. Or with a different pushing and pulling, the function could be transformed into a function having no critical points at all. Similarly, if two critical values are the same, all one needs do to make the values different is to pull the graph down just a little bit near one of the critical points.

With these intuitive geometrical ideas in mind, consider now a smooth function having only nondegenerate critical points. Such a function is called a *Morse function*. We then have

---

**MORSE'S THEOREM** *Morse functions are stable and dense in the set of all smooth functions. Furthermore, near a critical point of index* k *there is a smooth change of coordinates under which the resulting Taylor series of the Morse function* f *near the origin is the pure quadratic form*

$$f(x_1, x_2, \ldots, x_n) = -x_1^2 - x_2^2 - \cdots - x_k^2 + x_{k+1}^2 + x_{k+2}^2 + \cdots + x_n^2.$$

*In other words, under this coordinate change the Tayl of* f *vanishes.* ∎

The first part of the theorem says that a small, smooth perturbation of a Morse function yields another Morse function; thus, all functions sufficiently close to a Morse function are themselves Morse functions. The density result means that there is a Morse function arbitrarily close to any non-Morse function, that is, a function like $f(x) = x^3$ having a degenerate critical point. The last part of the theorem says that there is a smooth change of coordinates near the critical point that transforms the function $f$ into a saddle, and that the exact nature of the saddle is determined solely by the quantity $k$, the index of the critical point.

The foregoing local stability result for Morse functions can be upgraded to a *global* result by imposing the additional condition that the critical values be distinct. A Morse function $f$ satisfying this condition is globally stable, which means that a small, smooth perturbation of $f$ defined over the *entire* space $R^n$ yields another Morse function with distinct critical values. This is in contrast to the case of a *local* perturbation, which may be defined only over a small region in $R^n$ containing a single critical point, as was illustrated earlier in Figure 3.8.

At first glance, it may seem that Morse's Theorem closes out the problem of what we can expect from smooth functions since the theorem tells us that any function can be approximated as closely as we wish by a Morse function. The theorem even gives a complete picture for what every Morse function looks like near its critical points. And we know that this is all the information needed to reconstruct the overall behavior of the function. So, in this sense the theorem does indeed close out the issue of the local behavior of smooth functions. But in another sense it is only the beginning of the story, not the end, encouraging us to ask how Morse's Theorem might be extended to even more general situations. As it turns out, there are several interesting directions to look for such extensions.

- *Smooth maps:* Morse's Theorem applies only to smooth *functions,* transformations whose values are simply real numbers. But what about situations in which we have a smooth *map,* as in the case of Whitney's Theorem, where the map took its values in the two-dimensional plane $R^2$ instead of the reals? Even more generally, what if the maps take their values in some higher-dimensional space like $R^p$ for $p$ greater than 2? The fact that such maps take on values in a far bigger space than just the real line opens up possibilities for much more varied behaviors than mere functions can display. Whitney's Theorem, which we outlined in the opening section, gives a pretty complete picture of what can happen for maps of the plane to the plane. But a decent general theory for smooth maps of $R^n$ to $R^p$ remains but a dream for $p$ larger than 2.

- *Degenerate critical points:* Morse's Theorem also applies only to functions that are Morse functions—functions that by definition have only nondegenerate critical points. Since we've seen that a degenerate critical point can always be transformed into a non-degenerate one by a small "jiggling" of the original function, it may seem that there is no compelling reason to devote attention to these unstable cases; they can always be perturbed away using a perturbation as small as we like. This is all true. But there are many ways in which a function can be perturbed, an infinite variety of them, in fact. So it's of interest to ask about the *simplest* kind of perturbation that will remove the degeneracy. And even more, we might ask about the simplest *family* of functions into which our degenerate member can be embedded, so that the family as a whole retains its properties if we jiggle it a little bit into a new family. In other words, what we are looking for is an entire family of functions within which we can "hide" the degenerate member so that the family itself is stable (this sounds a bit like some third-rate television soap opera, doesn't it?).

To illustrate the problem here, consider a predator-prey ecosystem in which the predator population is $x$ and that of the prey is $y$. Suppose these populations depend on the prey birth rate, which we'll call $a$, and the predation rate, which we shall label $b$. Then, the total biomass in the system might be described by the function $V(x, y; a, b)$, which we can assume has a critical point at the origin. By fixing values for $a$ and $b$, we pin down a specific model for the biomass. But for certain values of these parameters, the critical point at the origin goes from being nondegenerate to degenerate. And this changes the local behavior of $V$—perhaps dramatically. It's of considerable interest to know exactly *how* changes in the biomass take place as $a$ and $b$ pass through these sensitive values. This is where the idea of placing the function $V$ into a family of smooth functions comes into play.

Let's look at the situation geometrically, considering again the function $f(x) = \frac{1}{3}x^3$. We may regard $f$ as a member of the family of functions $f_a(x) = \frac{1}{3}x^3 + ax$, where the real number $a$ serves to label or "name" the members of the family. This family is shown in Figure 3.11. So the original degenerate critical point occurs in the family member whose "name" is $a = 0$, all other family members having only a pair of nondegenerate critical points at $x = \pm\sqrt{-a}$ when $a$ is negative and no critical points at all for positive values of $a$.

No smooth perturbation of the family as a whole will remove the degeneracy from the family. Any new family arising from such a perturbation will have a member that displays exactly this same type of degenerate critical point. But as we'll soon see, the methods of singularity theory assure us that the family $\frac{1}{3}x^3 + ax$ is stable *as a family*. This means that if we move to a nearby family of functions, the two families will contain functions with the same critical point structures. Moreover, this is the simplest family (in the sense of requiring the smallest number

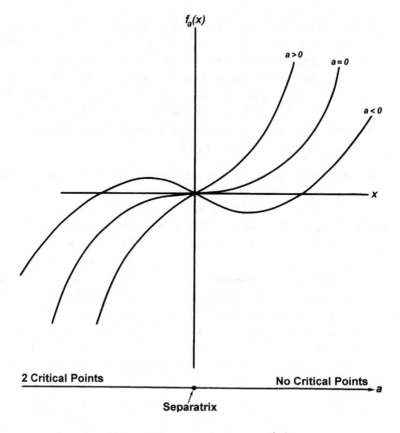

**Figure 3.11**  The family of functions $\frac{1}{3}x^3 + ax$.

of constants to name its members, specifically 1) that will "take-in" the unstable element $\frac{1}{3}x^3$. From this we see that a family can be stable *as a family,* even though it may contain a "black sheep" family member that is unstable *as an individual.*

The situation just described is one that comes up regularly in modeling processes throughout all areas of science, since function families arise naturally whenever the process under investigation contains any constants (technically, parameters) whose values need to be set to completely specify the process. So, for instance, when we consider the flow in a chemical reactor or a predator–prey system in ecology there are quantities in the model like reaction constants or predator birth and death rates that can vary. By fixing values for these parameters, we single out a particular member of a family of models. And if there is even one member of this family that has a degenerate critical point, which might correspond to a point where an ecological food web breaks down or where the flow of a fluid becomes turbulent, it will not be possible to eliminate this degeneracy by perturbing the family. Instead, we have to deal with how the behavior of the system changes as we move through the family member containing this "bad" point via nearby "nice" members. The generalization of Morse's Theorem enabling us to do this is a branch of singularity theory that in the popular-science press has come to be termed *catastrophe theory.*

The goal of catastrophe theory is to classify smooth functions with degenerate critical points, just as Morse's Theorem gives us a complete classification for Morse functions. The difficulty, of course, is that there are a lot more ways for critical points to "go bad" than there are for them to stay "nice." Thus, the classification problem is much harder for functions having degenerate critical points, and has not yet been fully carried out for all possible types of degeneracies. Fortunately, though, we can obtain a partial classification for those functions having critical points that are not too bad. And this classification turns out to be sufficient to apply the results to a wide range of phenomena like the predator–prey situation sketched above, in which "jumps" in the system's biomass can occur when parameters describing the process change only slightly.

This classification is important since it tells us the different ways by which a nondegenerate critical point can become degenerate as we vary parameters describing a function family. Such information, in turn, allows us to identify those combinations of parameter values at which the system can show discontinuous shifts—crashes, outbreaks, collapses— in its behavior. This is the applied value of catastrophe theory. So let's see how it works.

## The Thom Classification Theorem

Our concern now is with the behavior of smooth functions that are not Morse functions, what we might term the "bad" smooth functions, those with degenerate critical points. The key to the problem of classifying the types of behaviors such a function can display lies in figuring out a way to measure just how bad a function actually is. Basically, there are two different—but not completely independent—ways of measuring a function's badness. The first involves determining the degree of the degeneracy of the critical point. This measure, termed the *corank* of the function, is a condition on the second-order terms in the Taylor series expansion of the function. There are $n(n+1)/2$ such terms for functions of $n$ variables, and the corank measures the degree to which these terms are independent of each other. The more independence among the terms, the less the degeneracy of the critical point.

The second measure of badness quantifies how far away the function is from being a Morse function. This measure is called the *codimension,* and is an integer computed from the first-order terms of the function's Taylor expansion. There are $n$ such terms for functions of $n$ variables, with the codimension being related to the degree of dependency among these quantities.

Recall that for a Morse function of $n$ variables, the index $k$ of a critical point is simply the number of coordinate directions in which the function's local curvature is negative, while the complementary quantity $n - k$ measures the number of directions of positive local curvature. In the case of a function of a single variable ($n = 1$), the vanishing of the second-order expansion coefficient in the Taylor series at a point means simply that the function is locally "flat" at that point. By analogy, for a function of $n$ variables the corank measures the number of coordinate directions in which the function is flat. In particular, for a Morse function there can be no such directions since the critical points are all nondegenerate. This translates into the algebraic fact that the corank is 0 at such critical points. And, of course, the more flat directions there are, the greater is the degeneracy of the critical point. So this is the way the corank measures the degeneracy of the critical point. Now what about the codimension?

Geometrically, we can regard the space of smooth functions as an infinite-dimensional space whose coordinate directions near the origin can be labeled $1, x, x^2, x^3, \ldots$. So if we have a smooth function that is not Morse, we can ask about the minimal number of independent terms (that is, "directions" in the space of smooth functions) that need to be added to the non-Morse function to make it Morse. This number is the codimension. Another way to think about this is to look for the simplest stable family of functions in which we can include (technically: embed) the non-Morse function. Here, again, "simplest" means the family with the smallest number of parameters needed to describe it. The number of such parameters also turns out to equal the codimension.

To illustrate these notions, consider still once again the simplest non-Morse function $f(x) = \frac{1}{3}x^3$. (Recall: by definition, a non-Morse function must have no linear or quadratic terms.) We have seen that this "bad" function can be embedded within the stable one-parameter family of functions $f_a(x) = \frac{1}{3}x^3 + ax$. So $f$ has codimension 1, since we need only to add the single term $ax$ in order to create a stable family within which the original function can be embedded. Moreover, it's easy to see that the corank of the critical point at the origin is also 1, since a function of a single variable can have at most one flat direction (which this one does).

Happily, the mathematical universe is arranged so that these two quantities—corank and codimension—are all we need to know in order to characterize and describe completely non-Morse functions that are not too bad. What the Thom Classification Theorem accomplishes is to show that up to a smooth change of variables, there are only a small, *finite* number of nonequivalent, non-Morse functions, and that each of these "standard" non-Morse functions can be described just by knowing the corank of the function. In addition, Thom's result gives us the explicit form of the simplest stable family into which the non-Morse function can be placed, this family being described in terms of the codimension of the original non-Morse function.

This pivotal result was first conjectured by René Thom in the 1960s, a complete proof being given later by John Mather, based upon key work of Bernard Malgrange. The Russian mathematician V. I. Arnold has substantially extended the original results, and the implications of the Thom Classification Theorem for applications have been extensively pursued by Christopher Zeeman, Tim Poston, Ian Stewart, Michael Berry, and many others. So, in a very real sense, the Thom Classification Theorem, while rightly attributed to Thom, is the product of a large group of mathematicians and represents a classic example of how fundamental mathematical advances are put together, piece by piece, in a collective international effort. Now let's give a statement of the result itself.

Recall that Morse's Theorem told us that all functions were equivalent to a pure quadratic near a nondegenerate critical point. Clearly, since quadratic functions by their very nature fall into the category of functions without a degenerate critical point, only functions of order cubic and higher can possibly have such "bad" critical points. Thom discovered that if we restrict attention to functions of this type having codimension no greater than 4, then it's possible to classify this subset of the space of all smooth, non-Morse functions. In addition, the codimension and the corank taken together serve to label the nonequivalent, non-Morse functions and their stable families, just as the index $k$ labeled the different types of Morse functions. Here is a full statement of the basic result.

---

**THOM CLASSIFICATION THEOREM**   *Up to multiplication by a constant and addition of a nondegenerate quadratic form, every non-Morse function of codimension less than or equal to 4 is smoothly equivalent near the origin to one of the standard forms given in Table 3.1.*   ∎

Probably the first thing the reader will notice about the entries in Table 3.1 is the column headed "Function," which lists only functions of one or two variables. This may seem a bit odd since we are supposed to be dealing with functions of $n$ variables. The reason for this oddity is a mathematical fact called the *Splitting Lemma*. It says that by a smooth change from the $x$ variables to a new set of variables $y$, the original non-Morse function $f$ "splits" into two parts, so that we can write $f(y)$ as $f(y) = p(y) + Q(y)$. Furthermore, the number of $y$ variables appearing in the function $p$ equals the corank of the original function $f$, while the remaining variables are all in the $Q$ part of the splitting. Finally, the Splitting Lemma says that $p$ is a function of order cubic or higher, while $Q$ is a pure quadratic and thus contains none of the degeneracy of the original function $f$. So by changing to the new variables $y$, all the degeneracy making $f$ a non-Morse function is squeezed into the first part of the splitting, the function $p(y)$. So for ease of writing, Table 3.1 shows only this degenerate part of $f$ and omits the quadratic part $Q$, which the reader should mentally tack on in each case.

The column labeled "Universal Unfolding" in Table 3.1 shows the simplest function family that is both stable *as a family,* and which contains the non-Morse function as a member. These families, which are called *unfoldings,* are labeled by the real numbers $a_1$, $a_2$, and so forth. From Table 3.1, we see that the number of parameters in a minimal-

**Table 3.1** The Thom classification of smooth functions.

| Corank/ Codimension | Function | Universal Unfolding | Name |
|---|---|---|---|
| 1/1 | $y^3$ | $y^3 + a_1 y$ | Fold |
| 1/2 | $y^4$ | $y^4 + a_1 y^2 + a_2 y$ | Cusp |
| 1/3 | $y^5$ | $y^5 + a_1 y^3 + a_2 y^2 + a_3 y$ | Swallowtail |
| 1/4 | $y^6$ | $y^6 + a_1 y^4 + a_2 y^3 + a_3 y^2$ $+ a_4 y$ | Butterfly |
| 2/3 | $y_1^3 - 3y_1 y_2^2$ | $y_1^3 - 3y_1 x_2^2 + a_1(y_1^2 + y_2^2)$ $+ a_2 y_1 + a_3 y_2$ | Elliptic umbilic |
| 2/3 | $y_1^3 + y_2^3$ | $y_1^3 + y_2^3 + a_1 y_1 y_2 + a_2 y_1$ $+ a_3 y_2$ | Hyperbolic umbilic |
| 2/4 | $y_1^2 y_2 + y_2^4$ | $y_1^2 y_2 + y_2^4 + a_1 y_1^2 + a_2 y_2^2$ $+ a_3 y_1 + a_4 y_2$ | Parabolic umbilic |

parameter family equals the codimension of the function. The names in the final column of Table 3.1 come from the shapes the families assume when graphed as a function of the parameters $a_i$. It's what happens when we move through one of these families by varying these parameters that is the proper subject of "applied" catastrophe theory, a matter to which we shall turn now.

Incidentally, if you're wondering why the Thom Classification Theorem stops after functions of codimension 4, the reason is that when the codimension gets much larger than this the classification is no longer finite; there are then an infinite number of equivalence classes. However, it is the finite, low-codimension classification that is the most important for applications. And we can regard it as a gift from the mathematical gods that such a classification exists, at all, even for functions of low codimensions.

Probably the most important reason that catastrophe theory received as much popular press as it did in the mid-1970s is not because of its unchallenged mathematical elegance, but because it appears to offer a coherent mathematical framework within which to talk about how discontinuous behaviors—stock market booms and busts or cellular differentiation, for instance—might emerge as the result of smooth changes in the inputs to a system, things like interest rates in a speculative market or the diffusion rate of chemicals in a developing embryo. These kinds of changes are often termed *bifurcations,* and play a central role in applied mathematical modeling. Catastrophe theory enables us to understand more clearly how—and why—they occur.

The reason catastrophe theory can tell us about such abrupt changes in a system's behavior is that we usually observe a dynamical system when it's at or near its steady-state, or equilibrium, position. And under various assumptions about the nature of the system's dynamical law of motion, the set of all possible equilibrium states is simply the set of critical points of a smooth function closely related to the system dynamics. When these critical points are nondegenerate, Morse's Theorem applies. But it is exactly when they become degenerate that the system can move sharply from one equilibrium position to another. The Thom Classification Theorem tells when such shifts will occur and what direction they will take. This claim will now be illustrated with an important example from engineering mechanics.

## Bridges and Beams

In gale winds of over 40 miles per hour, the Tacoma Narrows Bridge, which linked the mainland of the state of Washington to the Olympic Peninsula, came crashing down into Puget Sound on the morning of November 7, 1940—less than 4 months after its opening. With this collapse, the bridge gained instant immortality (and notoriety) as one of history's great engineering failures. Postmortems on the bridge's collapse showed that the 2,800-foot main span went into a series of torsional oscillations, whose amplitudes increased until the convolutions tore several of the suspension cables loose, at which point the span broke up and fell into the water. The reader can get a feel for the forces at play in this famous disaster from Figure 3.12, which shows the span twisting like a demented snake shortly before breaking up. While the Tacoma Narrows Bridge disaster was attributable to dynamic resonance, a bifurcation in a *dynamical* process, catastrophe theory can help us analyze similar, but *static,* situations involving the bending, breaking, and collapsing of columnar beams.

Suppose we have an elastic strut of length $\ell$ subject to a force $K$ exerted at each of its ends. The inherent symmetry in this situation will be destroyed due to manufacturing imperfections, causing the beam to buckle either upward or downward as $K$ is increased. If the beam

**Figure 3.12** The Tacoma Narrows Bridge collapse.

buckles upward, and we agree to measure the amount of buckling by the quantity $x$, we end up with the situation shown in Figure 3.13.

Now assume that a load $L$ is applied to the center of the strut. Then the displacement $x$ will decrease continuously until the load reaches a critical value, at which point the strut will suddenly jump from being buckled upward to a downward-buckled state. Catastrophe theory enables us to study these discontinuous shifts as we smoothly vary the two parameters $K$ and $L$. Figure 3.14 shows the various possibilities.

Initially, the beam is in an unbuckled state ($x = 0$). As the two parameters change along the numbered path in Figure 3.14, the strut behaves in the following manner: Nothing happens as $K$ increases until

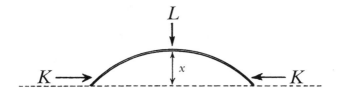

**Figure 3.13** An elastic strut.

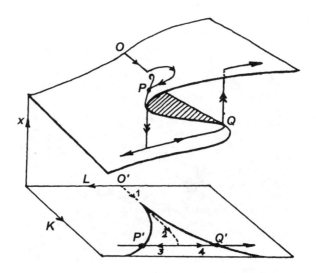

**Figure 3.14** Geometry of beam buckling.

the cusp point is reached, whereupon the strut begins to buckle upward. More than 200 years ago, the Swiss mathematician Leonhard Euler showed that this occurs when $K = \pi^2 \lambda / \ell^2$, the quantity $\lambda$ serving to measure the elasticity of the strut. As $K$ is further increased along path 2, the strut buckles more sharply. If we now keep $K$ constant and increase $L$ along path 3, the displacement gradually decreases until at the point $P'$ it jumps onto the other sheet of the behavior surface, which corresponds to a sudden snap into the downward-buckled state. If the load $L$ is decreased toward point 4, the amount of buckling changes continuously, and the strut maintains its downward-buckled state past the point $P'$. In other words, it does not return to the upper surface until it reaches $Q'$, at which point it suddenly snaps into the upward-buckled state.

You might well ask: How do we know that the geometry shown in Figure 3.14 is the right one for this situation? The answer ultimately rests with the precise mathematical form of the function characterizing the total energy of the beam. Readers interested in the details of this energy function are invited to consult the material cited in the bibliography. Classical physics tells us that the beam always seeks to move to a state that minimizes its total energy. Since the expression for the energy defines a *family* of functions, each member of the family being identified by a particular value of the two parameters $K$ and $L$, we have a two-parameter family. We will now see how the Thom Classification Theorem allows us to conclude that this is the standard family for a function of codimension 2 and corank 1. This, in turn, will allow us to assert that the geometry shown in Figure 3.14 is indeed the one governing the behavior of the beam.

# Bifurcations, Catastrophes, and Equilibria

The function family $f_a(x) = \frac{1}{3}x^3 + ax$ displayed above in Figure 3.11 shows that as we move through the family by smoothly changing the parameter $a$, we move from functions having two nondegenerate critical points to functions having no critical points, the dividing line being the function $\frac{1}{3}x^3$ which has a single degenerate critical point at the origin. The parameter value(s) at which the critical point structure of the function family changes is called a *bifurcation point*. Such points are particularly important for applications, since a change in the critical point structure usually corresponds to some dramatic shift in the physical process the function represents. Part (a) of Figure 3.15 shows the above cubic family, while Part (b) displays the *bifurcation diagram* for this function family. The upper and lower branches of the curve for negative values of $a$ in Part (b) of the figure are the locations of the nondegenerate critical points as a function of the parameter $a$. It is the shape of this curve that gives rise to the term "fold" for this type of degenerate critical point. From the diagram, we see how the two nondegenerate critical points merge into a single degenerate critical point at $a = 0$, which in turn gives way to no critical points, at all, as we move on through the function family via positive values of $a$.

The situation gets even more interesting when we consider two-parameter families. Table 3.1 showed that the standard family of this type is the family of quartics $\frac{1}{4}x^4 + \frac{1}{2}a_1x^2 + a_2x$. The bifurcation diagram for this family is shown in Figure 3.16. The convoluted surface in the upper part of the figure plays the same role for this family as the two curves in Figure 3.15, in that each point on the surface represents a critical point of the function family for the corresponding value of the parameters $a_1$ and $a_2$. Of great importance is the fact that the set of points in the two-dimensional parameter space at which there is a change in the number of

119

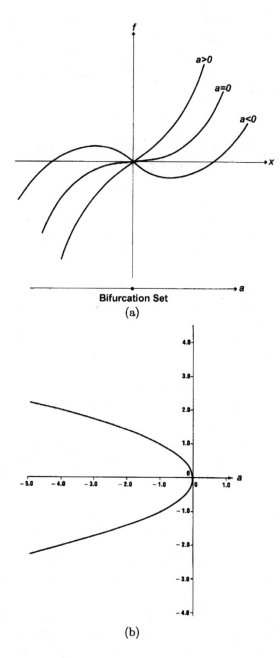

**Figure 3.15**    Bifurcation diagram for the family $\frac{1}{3}x^3 + ax$.

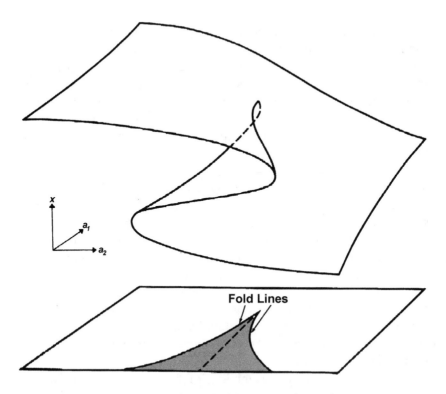

**Figure 3.16**   Bifurcation diagram for the family $\frac{1}{4}x^4 + \frac{1}{2}a_1x^2 + a_2x$.

critical points is now a pair of curves—the fold lines—rather than just a single point, as was the case in the one-parameter family discussed above. As noted in the figure, the functions have three nondegenerate critical points for all parameter values lying inside the shaded region bounded by the two fold lines. On the fold lines themselves, two of the critical points have merged into a single degenerate critical point, while at the cusp point, where the fold lines come together, there is a single degenerate critical point of corank 2 instead of 1. Finally, outside the shaded region, the functions in the family have a single nondegenerate critical point. Again, the name "cusp" for this kind of degenerate critical point of codimension 2 comes from the geometry of the bifurcation lines in the space of parameters $a_1$ and $a_2$.

Comparing Figures 3.15 and 3.16, the perceptive reader might note a striking similarity between the curve of Figure 3.15 and the curve that would be obtained by slicing the surface in Figure 3.16 along a fixed value of $a_1 < 0$. In fact, the diagram of Figure 3.15 is just a subdiagram of Figure 3.16. And this is true for all the functions of a given corank: Each bifurcation surface is contained as a proper subset of those for functions of the same corank and higher codimension. The corresponding diagrams for families involving three or more parameters are a bit more difficult to draw in their entirety, but there is no problem in writing down their algebraic expressions. When we do, it turns out that the bifurcation diagram for such families contains all the types of bifurcations seen for function families of fewer parameters—as well as a lot of new bifurcation possibilities opened up by the additional degrees of freedom created by the higher codimension. Readers interested in seeing the geometry of sections of these higher-dimensional surfaces should consult the material cited for this chapter in the bibliography.

The points, curves, and surfaces in the space of parameters at which the critical point structure changes are called *catastrophe points,* a terminology that stems mostly from attaching a physical interpretation to the critical points. On this note we finally arrive at the juncture in our narrative where it's time to direct attention to why these singularities and their bifurcation structures are so important both in natural and human affairs.

Up to now, we have been thinking of the critical points of a smooth function mostly in mathematical terms, merely as places where the function has a local minimum, maximum, or something more complicated—in short, as points where "something funny" happens in the behavior of the function. But there is another way to view these points, one more congenial to physicists and biologists than to mathematicians, a view showing how singularity theory makes contact with dynamical system theory. This is to regard the critical points as the equilibrium points of a dynamical process, that is, the points at which the dynamical process comes to a dead stop and sits forever unless perturbed away from this point by forces outside the system. Of course, only particular types of dynamical processes admit this kind of identification of their long-run behavior, namely, those processes having only points as their long-run behavior and not more complicated structures like periodic orbits or chaotic trajectories. In particular, we consider here only those dynamical processes whose equilibrium points coincide with the critical points of a smooth function $f$. Imposing this condition allows us to use singularity theory to analyze how the system's equilibrium state changes as we vary the parameters describing the system. Such dynamical processes are called *gradient systems.*

Initially, it may seem that such systems constitute a very special class of processes. And, in fact, that is indeed the case. However, nature has providentially worked things out so that a lot of processes of practical concern just happen to belong to this class—including many of the systems of classical physics like passive electrical circuits, damped vibrating springs, and bending beams. Moreover, when we observe these kinds of processes in real life, what we usually see is the system when it is at or very near to equilibrium. For these reasons catastrophe theory can be of great value in helping us understand how these kinds of systems can shift abruptly from one equilibrium state to another as various parameters, like spring constants or unemployment rates, are varied just a little bit. So we will now shift our attention away from purely mathematical considerations, and look at some examples of this kind of applied analysis in action. But first, a small caveat.

## How Local Is Local?

Catastrophe theory is a *local* theory, telling us what a function looks like in a small neighborhood of a critical point; it says nothing about what the function may be doing far away from the singularity. Yet most of the applications of the theory—including those to follow in this chapter—involve extrapolating these rock-solid, local results to regions that may well be distant in time and space from the singularity. So how local is local? How far away from the critical point can we move before the local structure provided by the Thom Classification Theorem no longer applies? In general, there is no clear-cut mathematical response. To answer it satisfactorily, the mathematical modeler has to take refuge in the *philosophy* of modeling rather than in its mathematics.

When we examine the modeling literature, its most striking aspect is the predominance of "flat" linear models. Why is this the case? After all, from a singularity theory viewpoint these linear objects are mathematical rarities. On mathematical grounds we should certainly not expect to see them put forth as credible representations of reality. Yet they are. And the reason is simple: linearity is a neutral assumption that leads to mathematically tractable models. So unless there is good reason to do otherwise, why not use a linear model?

123

When it comes to modeling processes that are manifestly governed by nonlinear relationships among the system components, we can appeal to the same general idea. Calculus tells us that we should expect most systems to be "locally" flat; that is, locally linear. So a conservative modeler would try to extend the word "local" to hold for the region of interest and would take this extension seriously until it was shown to be no longer valid.

So, even though we know that the behavior of a function expressed by the Thom Classification Theorem near a critical point does not hold far away from that point, we blithely assume that it does—until the behavior of the system shows otherwise. And when it does show otherwise, we have gained information about where this neutral assumption fails. Moreover, this then enables us to accumulate evidence and information about how to model the situation more accurately. The reader should keep this point uppermost in mind as we make our way through the applications of catastrophe theory that follow.

The key ingredient making the catastrophe theory magic work in the collapsing beam example shown above is the fact that centuries of work in mechanics has given us a very good understanding of the precise mathematical form of the energy function governing the behavior of a beam. Having this function in our hip pocket, so to speak, it's a simple matter to pull the relevant mathematical tools from singularity theory off the shelf and apply them. When we did this for the bending beam, what popped out was the fact that the cusp catastrophe was the right way to describe the ways the beam can buckle. But there are other situations in which catastrophe-theoretic arguments are used, where we do not have such complete knowledge of the function governing the system dynamics. A good example of this kind of analysis arises in developmental biology.

## The Shape of Things

The greatest unsolved problem in biology, the real terra incognita of the field, is the enigma of embryology. How is it that an initially homogeneous ball of cells can differentiate and organize itself into one of the myriad species of living things we see making their living on Earth today? This is the problem of cellular differentiation and morphogenesis, or more prosaically, the emergence of form. Stimulated by work on morphogenesis in the late 1950s by the British developmental biologist C. H. Waddington, René Thom originally developed catastrophe theory as a mathematical way of trying to address this very question.

The big question Thom wanted to answer mathematically is how the basic topology of an organism is specified. Morphogenesis doesn't take place simply by cells of the right type forming in appropriate positions in the growing embryo. In fact, early on in the embryonic process cellular types are not usually determined at all. Rather, most cells are capable of ultimately developing into one of a number of different cellular types for some period of time after their initial formation. So what is it that causes a general-purpose type of cell to suddenly become a liver cell or a muscle cell or a brain cell? And once its fate is decided, how does the cell know where to move and when in order to take its proper place in the final adult organism? In a nutshell, these questions constitute the mystery of morphogenesis.

Nobody really knows how the fate of an individual cell is decided. What is clear, however, is that neighboring cells interact and that there are chemical gradients within the growing embryo. One of the first researchers to propose an explicit mathematical model to explain development based on these chemical gradients was, surprisingly enough, the computer scientist Alan Turing, who will play the starring role in our drama about the theory of computation in Chapter 4. In 1952, Turing proposed a mathematical model involving the processes of reaction and diffusion of various unspecified chemical compounds that he termed *morphogens*. Many investigators since that time have developed Turing's idea to a high degree of mathematical sophistication, leading to the view of some workers in developmental biology like Brian Goodwin and Hans Meinhardt that the ultimate fate of each cell is decided by the concentrations of various morphogens. These concentrations, in turn, are assumed to be determined by a dynamical process describing how the morphogens interact and change their levels and types.

Adopting Turing's basic idea, it may reasonably be supposed that there are time-dependent changes going on in the cells. To keep things as simple as possible, assume that the fate of each cell is decided by a single morphogen whose concentration is the equilibrium state of a dynamical process, which is just a single point geometrically. Such a state will, of course, depend on hundreds, or even thousands, of biochemical variables, which in turn generate the postulated morphogen. We can think of these biochemical quantities as being internal variables, inaccessible to observation. So the only observed output of the system is the single quantity, the morphogen concentration itself. Since there are only four independent directions in space and time in which we can look at the developing organism, we can think of this as a problem for which there are four independent input variables.

We now find ourselves in a familiar situation: a single output variable (the morphogen concentration) depending on four input variables (the spatial and temporal directions in which we can observe the developing embryo). We can think of these inputs as being knobs to twist, each setting specifying a particular location in space and time at which we observe the organism. Generally, the morphogen concentration, hence the observed properties of the organism, will vary smoothly as we slowly and continuously twist the knobs. However, more than one stable morphogen concentration will be possible for some knob settings. This results in the formation of "frontiers" between different types of tissues and between these tissues and the spatial region outside the organism (that is, the outside world). Thus, we conjecture along with Thom that in the local vicinity of a particular point in space and time, the physical form of the developing organism will be determined by one of the seven elementary catastrophes listed earlier in Table 3.1 (see the section "The Thom Classification Theorem," above).

It turns out that each of the "magnificent seven" catastrophe types listed in Table 3.1 can be given both a spatial and a temporal interpretation, depending upon whether we interpret the input variable(s) as space or time. Table 3.2 lists these possibilities. As an illustration of how to make use of these interpretations consider Figure 3.17, which shows a section of the parabolic umbilic catastrophe. The spatial interpretation of this catastrophe (from Table 3.2) suggests the formation of a mouth, which is indeed a good image of what seems to be emerging as we take different sections of this catastrophe surface. This figure is obtained by fixing two of the four parameters describing the parabolic umbilic family, and drawing the surface that results when we let the other two parameters vary. Here the resulting surface shows a striking resemblance between such a section of the parabolic umbilic and a bird's beak. Many other examples of this sort can be found in the material cited in the bibliography.

**Table 3.2**  Spatial and temporal interpretations of catastrophes.

| Catastrophe | Spatial Interpretation | Temporal Interpretation |
|---|---|---|
| Fold | Boundary | Beginning (ending) |
| Cusp | Pleat; fault | Separating (uniting); changing |
| Swallowtail | Split; furrow | Splitting; tearing |
| Butterfly | Pocket | Giving (receiving); filling (emptying) |
| Hyperbolic umbilic | Wave crest; arch | Collapsing; engulfing |
| Elliptic umbilic | Spike; hair | Drilling |
| Parabolic umbilic | Mouth | Opening (closing); ejecting |

Developmental biology is an area somewhere between physics and philosophy when it comes to being able to write down an explicit mathematical representation for the underlying processes we want to study. We know *something* about these morphological goings-on. But we don't know nearly as much as we know about phenomena like the bending beams studied above. This kind of analysis, where we have at least some knowledge of the underlying process, is what we might call the "semiphysical way" of catastrophe theory. This involves regarding the input variables as being the parameters characterizing a family of functions, while associating the output variables with the non-Morse part of whatever function it is that actually governs the process. We then use the physical knowledge we have available to constrain the relevant family of functions.

It's clear, I think, how very different this semiphysical way of catastrophe theory is from the physical way represented by our analysis of the bending beam. In that case, we knew everything about the

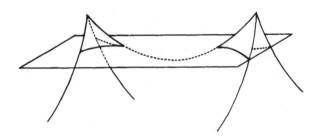

**Figure 3.17**  A section of the parabolic umbilic catastrophe.

127

functional form of the relationship between the variables of the problem. So it was a simple matter to just plug this mathematical expression into the machinery of singularity theory, turn the mathematical crank and eventually obtain the relevant bifurcation diagram. But the semiphysical way is not the end of the story. There remains what has come to be termed the "metaphysical way" of catastrophe theory, which turns up whenever we have no idea about the underlying dynamics. Let's look at how this way of employing catastrophe theory can be used to study humor.

## Laughs and Cries

In his immensely entertaining and enlightening book *Mathematics and Humor*, John Allen Paulos uses the cusp surface of catastrophe theory as a kind of mathematical metaphor for encapsulating in geometrical terms the essence of what makes a joke funny. Paulos begins his analysis by noting that a joke depends on the perception of incongruity in a given situation or its description. Thus, a joke can be regarded as a form of structured ambiguity between two (or more) possible interpretations. The joke's punch line then acts as the switch that moves the listener from one interpretation to another. Figure 3.18 provides a geometrical view of this kind of situation.

To illustrate the point, consider the tale of the young man who goes to a computer dating agency. He tells the counselor that he wants to meet someone who enjoys water sports, likes company, is comfortable in formal attire, and is very short. Integrating these various requirements, the computer matches him up with a penguin. Here the first interpretation that unfolds is that of a woman with a certain type of lifestyle (the upper sheet in Figure 3.18). The punch line involving the man being given a blind date with a penguin uncovers the second, hidden, interpretation, which can be geometrically regarded as a catastrophic drop from the upper sheet to the lower sheet.

The cusp metaphor for the structure of a joke can also be used to explore pictorially why small deviations in the beginning of a joke often result in the joke not being very funny. The geometrical reason for this is shown in Figure 3.19a, which can be thought of as saying that the story develops in such a way that its interpretation crosses on the wrong side of the ambiguity region, hence always remaining on the lower sheet of

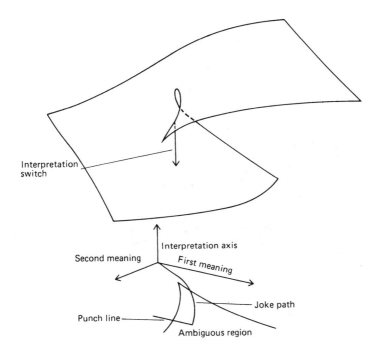

**Figure 3.18**   The cusp model of a joke.

the surface. Thus, the joke never has a chance to develop a buildup and ends up falling flat. On the other hand, Figure 3.19b shows the trajectory of a joke that's seen as being funny.

In his analysis of jokes as catastrophes, Paulos also notes how the shape of the catastrophe surface offers some insight into the importance of timing in the telling of jokes. A comedian must somehow sense how the audience has interpreted what he has already said, meaning that the comedian must know where the audience is located on the catastrophe surface. If the audience is ahead of the comedian, the alternate interpretation will become obvious too soon and the joke will lose its punch. On the other hand, if the comedian is ahead of the audience, the punch line will not precipitate the catastrophic jump in interpretation, leaving the comedian having to explain why the joke is funny.

The cusp geometry also suggests that the shift in interpretations brought about by the joke's punch line will be funnier if there is a large gap between the upper and lower sheets of the surface than if the gap is small. Paulos argues that this is likely to be the case for jokes that are

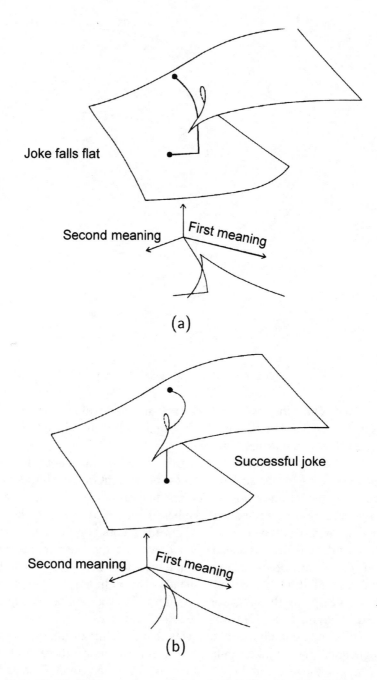

Joke falls flat

Second meaning   *First meaning*

(a)

Successful joke

Second meaning   *First meaning*

(b)

**Figure 3.19**   (a) The geometry of a joke that falls flat; (b) a joke that's funny.

about sex or authority, since people tend to have a greater than average level of anxiety about these topics.

While it's tempting to continue with a more extensive discussion of catastrophe theory as a"metaphysical" way to understand humor, I think the above sketch is sufficient to illustrate the fundamental ideas. Let's summarize what's been done.

One begins by singling out a handful of variables, calling them inputs. In the case of humor, these were the various meanings that could be given to the story. These quantities, in turn, are *postulated* to give rise to observable changes in another handful of quantities that we term outputs. In the case of jokes, the outputs can be the interpretations the listener gives to the story, or what might be even better, the physiological and mental tension and release brought on by the joke's story and eventual punch line. We further *posit* that there are no more than four inputs and two outputs. Next we *assume* that whatever the mathematical relationship linking the inputs to the outputs, the analytic structure of that relationship satisfies the technical conditions required by the Thom Classification Theorem. In broad terms, these conditions amount to the requirement that the system's outputs be the equilibrium points of some smooth dynamical process. Finally, we *presuppose* that the coordinate systems used to measure both the inputs and the outputs happen to coincide with the coordinate systems that lead to the standard catastrophe geometry governing a system with the number of inputs and outputs we have chosen. If not, we have to make a change of variables to transform the physically based coordinates in which we express the observations into the mathematically based variables that generate the standard geometries. The mathematics of catastrophe theory assures us that such a transformation exists—but gives no help in finding it.

Now focus your attention on the italicized words in the preceding paragraph. Each one of these words represents an out-of-the-blue assumption that we must be ready to swallow if we want to appeal to the Thom Classification Theorem to single out a particular geometry describing our problem. It's at this point that physics gives way to metaphysics. Or, put another way, it's at this point that modeling in the conventional sense with equations and variables with known interpretations gives way to Thomist-style *metamodeling,* which asks: If you use this *type* of equation, what sort of phenomena would you expect to see? Of course, the conventional wisdom in the modeling business regards the Thomist mode of thinking as being hopelessly speculative and just too philosophical to ever come to terms with the world as it is. This view, in fact, engendered a furious debate in the 1970s over the epistemological status of catastrophe theory, a debate that has by now happily run its course with neither side doing much by way of convincing the other of the merits of its case—the usual outcome of philosophical and religious arguments. Nevertheless, it's of historical interest, at least as a footnote in the sociopsychology of science, to conclude our discussion of singularity theory with a (very) brief reprise of some aspects of this mathematical brouhaha.

### Tempests and Teapots

In its issue of January 19, 1976, *Newsweek* magazine ran a full-page article on catastrophe theory, the first story on mathematics in the magazine in over 7 years. In this article the theory is described in terms rosy enough to emit heat, suggesting that Thom's ideas about discontinuous phenomena represent the most significant advance in applied mathematics since Newton's invention of the calculus. As we have come to expect whenever a laborer in the vineyards of mathematics receives even a smidgen of attention beyond the bounds of what the mathematical community feels is right and proper, the naysayers came out in force. In this case, the charge was led by Hector Sussman and Raphael Zahler, aided and abetted by a number of prominent colleagues. In a 1977 article by Gina Kolata in *Science* magazine titled "The Emperor Has No Clothes," the battle was joined in earnest, with several quotes on the demerits of catastrophe theory being uttered by prominent mathematicians like Stephen Smale of Berkeley and Joseph Keller of Stanford. In Kolata's notorious article, Christopher Zeeman, probably the most prominent advocate and expositor of "applied" catastrophe theory, is described as a "publicist," and Thom is rightly quoted, but completely out of context, to the effect that in a world in which all concepts could be formulated mathematically, only the mathematician would have a right to be intelligent. A long bout of correspondence, pro and con, followed on the issues raised in the article, little of which had any bearing on anything other than the attention that Thom and Zeeman were receiving from the world outside science and, especially, outside mathematics.

In trying to summarize and evaluate the pluses and minuses of the many threads in this debate, it's difficult for an uninvolved bystander not to wonder, Why all the fuss? As biologist Robert Rosen wisely counselled, "If an individual scientist finds such concepts uncongenial, let him not use them. There is no reason why he should take their existence as a personal affront." This is my view as well. Catastrophe theory will probably survive these broadsides, in much the same way and for much the same reasons that Darwin's theory of natural selection survived the bitter attacks mounted against it. Both theories are essentially explanatory rather than predictive, thereby failing to provide those who hunger for precise quantitative predictions with the kind of numerology that has come to be synonymous with *science*. But as René Thom so poignantly points out, "At a time when so many scholars in the world are calculating, is it not desirable that some, who can, dream?"

# CHAPTER

# 4

# The Halting Theorem

*Theory of Computation*

# Calculation versus Computation

The computer revolution began on a lazy English summer afternoon in 1935 when Alan Turing, an undergraduate at King's College, Cambridge, had the idea for a theoretical gadget by which to settle Hilbert's Decision Problem, a famous outstanding question in mathematical logic. At about the same time, heated debate in the Commons Room of the Princeton mathematics department over another mathematical tangle led to the development of a new kind of logical calculus, one that put the heuristic notion of what it means to carry out a computation on a sound mathematical footing. Taken together, these independent efforts by a handful of British and American mathematicians formed the theoretical foundation of what has come to be called "computer science." A decade later, motivated by his code-breaking work during World War II, Turing, along with John von Neumann and others in England and in the United States, began the process of translating these abstract mathematical notions about computation and logic into actual computing devices. The rest, of course, is history. This chapter examines the theoretical underpinnings of this work.

In 1935, Turing was sitting in on a set of lectures being given by the mathematical logician Max Newman. During the course Newman introduced Hilbert's *Entscheidungsproblem* (Decision Problem), which asks whether there exists an effective procedure for determining *in advance* if a certain conclusion follows logically from a given set of assumptions. Turing was immediately captivated by this unsolved conundrum, and took on the task of settling it. As it turned out, the central difficulty he encountered in trying to come to mathematical terms with the problem was that at the time there was no clear-cut notion of what was to count as an "effective procedure." Despite the fact that humans had been calculating for thousands of years, in 1935 there was still no good answer to the question, What is a computation? Turing overcame this difficulty by inventing a theoretical gadget that ended up serving as the keystone in the arch of the modern theory of computation.

Turing's primary task was to figure out how to replace the intuitively satisfying—but informal—idea of an effective process by a formal mathematical object. What he came up with is something we now call an *algorithm,* a notion he modeled on the steps a human being actually goes through when carrying out a computation. In essence, Turing saw an algorithm as a rote process, or a set of rules, telling one how to proceed under any particular set of circumstances. To illustrate the basic idea, think of the problem of making guacamole, a dip for tortilla chips popular in Mexico and parts of the American southwest.

## A Guacamole Algorithm

Guacamole is a combination of avocado, garlic, lemon juice, minced onion, chili powder, salt, and mayonnaise. Once these ingredients have been assembled, the recipe specifies in a step-by-step fashion how they are to be processed and combined to form the dip. For example, the first few steps might read:

1. Sprinkle the bowl with a little salt and rub with a clove of garlic.
2. Mash the avocado in the bowl and season with 1/4 teaspoon each of salt and chili powder.
3. Pour in 1 teaspoon of lemon juice.
4. Stir in 2 teaspoons of minced onions.

These steps constitute the initial part of what we might call a "guacamole algorithm." At each and every stage, the algorithm specifies an unambiguous action that's to be performed, culminating in a stopping rule telling us when the dip is complete. And, in fact, it's not hard to see that given the raw ingredients (and a sufficiently clever engineer), such a recipe could be followed in a purely mechanical fashion by a "guacamole machine."

Algorithms constitute the very core of the mathematical theory of computation. So now let's leave the kitchen and look at a less down-to-earth and more "down-to-mathematics" type of example.

## The Euclidean Algorithm

Consider the euclidean algorithm for finding the largest number dividing two given whole numbers $a$ and $b$. This is a rule for finding the greatest common divisor of the two numbers. In recent years the problem of finding the prime divisors of large integers has taken on increasing importance, since it bears directly on the issue of "uncrackable" schemes for encrypting data that's to be sent over open telephone lines. In particular, if we could guarantee that there were no good, that is, fast, ways to factor a large number into its prime divisors, then we could produce many such uncrackable codes. Consequently, algorithmic schemes for factoring numbers are of considerable research interest at the moment. The euclidean algorithm discussed below is well known to be computationally inefficient, that is, slow. Moreover, it only finds the greatest common divisor of the two numbers, which may or may not be a prime number. Nevertheless, the algorithm is of great theoretical interest in number theory and serves as a quintessential example of what we mean by an "algorithm."

Assume for the sake of definiteness that $a$ is larger than $b$. Moreover, introduce the notation "rem $\{\frac{x}{y}\}$" to denote the remainder after dividing a number $x$ by another number $y$. Then the euclidean algorithm consists of calculating the sequence of integers $\{r_1, r_2, \ldots\}$ by the rule

$$r_1 = \text{rem } \left\{\frac{a}{b}\right\}, \quad r_2 = \text{rem } \left\{\frac{b}{r_1}\right\}, \quad r_3 = \text{rem } \left\{\frac{r_1}{r_2}\right\}, \ldots,$$

where the process continues until a remainder of zero is obtained. The number $r^*$ at which the process halts turns out to be the largest integer dividing both $a$ and $b$. To explain exactly *why* this rule produces the greatest common divisor of $a$ and $b$ would take us a bit too far afield. But the reader interested in more details of this ancient algorithm can find the explanation in any book on elementary number theory.

To illustrate the process concretely, suppose we let $a = 137$ and $b = 6$. Then following the steps of the euclidean algorithm, we obtain

$$r_1 = \text{rem } \left\{\frac{137}{6}\right\} = 5, \quad r_2 = \text{rem } \left\{\frac{6}{5}\right\} = 1, \quad r_3 = \text{rem } \left\{\frac{5}{1}\right\} = 0.$$

We thus conclude that $r^* = 1$ is the greatest common divisor of 137 and 6. A perceptive reader will note that we didn't really need the euclidean algorithm to see this since 137 is a prime number; hence, the only number that divides 137 are 1 and itself. This then implies the greatest common divisor of 137 and 6 could only be 1. So 137 and 6 are what are termed *relatively prime* integers.

As with the guacamole example, what's important here is that the steps of the euclidean algorithm are rigidly prescribed, unvarying, and fixed in advance. One and only one operation is specified at each step, and there is no interpretation of the intermediate results or any skipping of steps—just a boring, basically mechanical repetition of the operations of division and keeping the remainder. This blind following of a set of rules is the essence of what constitutes an algorithm. Therefore, to reflect the mechanical nature of what's involved in carrying out the steps of an algorithm, Turing invented a hypothetical gadget now called a *Turing machine*. He then used the properties of this "machine" to formalize what it means to carry out a computation.

## Turing's Miraculous Machine

A Turing machine consists of two components: (1) an infinitely long tape ruled off into squares that can each contain one of a finite set of symbols, and (2) a scanning head that can read, write, and erase symbols from the squares on the tape. Since there is no gain in generality by allowing more than two symbols in the Turing machine alphabet, I'll assume henceforth that the symbols are just the two elements 0 and 1. For future reference, it's important to note that we're not thinking here of the symbols 0 and 1 as being the *numbers* zero and one, but only the *numerals* representing these numbers. And, in fact, we could just as easily have chosen any other two recognizably distinct symbols like the Roman numerals I and II, the letters $X$ and $Y$, or even the more abstract symbols ★ and ✠. However, for a variety of reasons, both historical and practical, it's convenient to adhere to the usual convention and use 0 and 1. The general setup for a Turing machine is depicted in Figure 4.1.

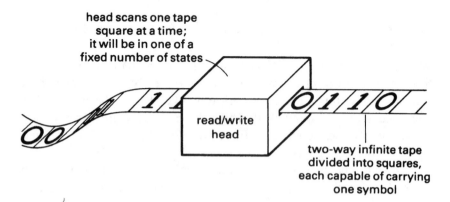

**Figure 4.1** Pictorial view of a Turing machine.

The behavior of the Turing machine is governed by an algorithm, which is manifested in what we now call a *program*. The program is composed of a finite number of instructions, each of which is selected from the following set of possibilities:

PRINT 0 ON THE CURRENT SQUARE
PRINT 1 ON THE CURRENT SQUARE
GO LEFT ONE SQUARE
GO RIGHT ONE SQUARE
GO TO STEP *i* IF THE CURRENT SQUARE CONTAINS 0
GO TO STEP *j* IF THE CURRENT SQUARE CONTAINS 1
STOP

That's it. From just these seven simple instructions we can compose what are called *Turing–Post programs*. These programs tell the machine what kind of computation it should carry out.

For future reference, let's note that the statements "GO TO . . . " in the above list of instructions can be eliminated by allowing the scanning head to have an internal mechanism, for example, a pointer, that can exist in any one of a finite number of positions A, B, C, and so on at each step of the computation. We call these different pointer configurations the *internal states* of the scanning head. Introduction of internal states substantially shortens the written expression of Turing machine programs, but in no way changes the kinds of computations that the machine can carry out.

The operation of the Turing machine is simplicity itself. We first feed in a tape containing a certain pattern of 0s and 1s (the input data). The machine then begins by placing the scanning head at some agreed-upon starting square. Thereafter, the actions taken by the machine are completely governed by the instructions in its program. But rather than continuing to speak in these abstract terms, it's simpler to just run through an example in order to get the gist of how such a device works.

Suppose the initial tape configuration consists of a string of 1s with a 0 at each end like this:

| · · · | 0 | **1** | 1 | 1 | 1 | 1 | 1 | 1 | 1 | 1 | 0 | · · · |
|-------|---|-------|---|---|---|---|---|---|---|---|---|-------|

In the tape configuration above, the boldface entry indicates the square at which the scanning head is currently located. For this example, assume that we want the Turing machine to change the 0s at the ends of the string into 1s and then stop, thus increasing the length of the block of 1s by two. Here is a program that accomplishes this task:

1. GO RIGHT ONE SQUARE
2. GO TO STEP 1 IF THE CURRENT SQUARE CONTAINS A 1
3. PRINT 1 ON THE CURRENT SQUARE
4. GO LEFT ONE SQUARE
5. GO TO STEP 4 IF THE CURRENT SQUARE CONTAINS A 1
6. PRINT 1 ON THE CURRENT SQUARE
7. STOP

By tracing through the steps of this simple program, we find that the scanning head moves right until it finds the first 0, which it replaces with a 1 by the PRINT 1 command. The head then moves left until it finds a 0 and replaces that with a 1, whereupon it halts. Figure 4.2 gives a more diagrammatic view of the general manner in which a Turing machine works.

Modern computing devices, even home computers like the one I'm using to write this book, seem to be vastly more complicated in structure and far more powerful in their computational capabilities than a Turing machine with its handful of internal states and very circum-

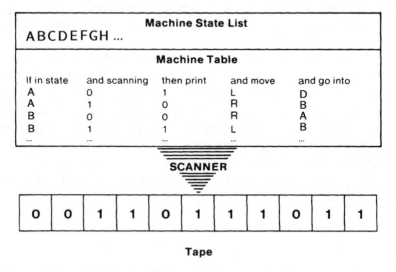

Figure 4.2 Diagrammatic representation of a Turing machine.

scribed repertoire of scanning head actions. Nevertheless, this turns out not to be the case. And a large measure of credit is due to Turing for recognizing that *any* algorithm (that is, program) executable on *any* computing machine—idealized or otherwise—can also be carried out using a particular version of his machine, termed a *universal Turing machine* (or UTM for short). So except for the speed of the computation, which definitely *is* hardware dependent, there's no computation that my machine (or anyone else's) can do that can't be done with a UTM—given enough time and memory.

To specify his UTM, Turing realized that not only the input data of the problem but also the program itself could be coded by a series of 0s and 1s. Table 4.1 shows one of the many ways this coding can be done. Consequently, we can regard the program as another kind of input data, writing it onto the tape along with the data it is to process. With this key insight at hand, Turing constructed a program that could simulate the action of any other program $\mathcal{P}$ when given $\mathcal{P}$ as part of its input (that is, he created a UTM). Let's consider how it works.

Suppose we have a Turing machine program $\mathcal{P}$, which we now know completely specifies a particular Turing machine. Thus, all we need do is write this program $\mathcal{P}$ onto the tape of the UTM, along with the input data that the program $\mathcal{P}$ is to act upon. Thereafter the UTM will simulate the action of $\mathcal{P}$ on the data; there will be no recognizable

**Table 4.1**   A coding scheme for the Turing machine language.

| Program Statement | Code |
|---|---|
| PRINT 0 ON THE CURRENT SQUARE | 000 |
| PRINT 1 ON THE CURRENT SQUARE | 001 |
| GO RIGHT ONE SQUARE | 010 |
| GO LEFT ONE SQUARE | 011 |
| GO TO STEP *i* IF THE CURRENT SQUARE CONTAINS 0 | $10\underbrace{100\ldots01}_{i \text{ repetitions}}$ |
| GO TO STEP *j* IF THE CURRENT SQUARE CONTAINS 1 | $11\underbrace{011\ldots10}_{j \text{ repetitions}}$ |
| STOP | 100 |

143

difference between running the program $\mathcal{P}$ on the original machine or having the UTM simulate $\mathcal{P}$, that is, pretend it *is* the Turing machine $\mathcal{P}$.

What's important about the Turing machine from a theoretical point of view is that it represents a formal mathematical object. Thus, with the invention of the Turing machine, for the first time there existed a well-defined notion of what it means to compute something. But this then raises the question of what exactly can be computed with such a mathematical gadget. In particular, is there a suitable Turing machine that will compute every number? Or do there exist numbers that are forever beyond the bounds of computation? Turing himself addressed this problem of computability in his trail-blazing 1936 paper, in which he introduced the Turing machine as a way of answering these fundamental questions.

## The Uncomputable

To begin our discussion of computability, let's first be clear on what we mean by a number being computable. Put simply, an integer $n$ is defined to be *computable* if there is a Turing machine that will produce it. That is, a number $n$ is computable if starting with a tape containing all 0s, there exists a Turing machine program that will stop after a finite number of steps with the tape then containing a string of $n$ 1s and all the rest 0s. Note that this is a *definition* of what it means for a number to be computable, and depends on this Turing machine model of computation. This implies that a particular number might be computable under the Turing machine model but uncomputable under a different mathematical idealization of what it means to compute something. This matter will be taken up in more detail at the end of the chapter. But for now it's important just to note that the concept of a number being computable is not an absolute notion; it depends on the model of computation you are using.

The case of computing a real number is a bit trickier since most real numbers consist of an infinite number of digits. For example, numbers like $\pi = 3.14159265\ldots$ or $\sqrt{3} = 1.732\ldots$ continue on forever, never settling into any repetitive finite cycle of numbers. So we call a real number computable if there is a Turing machine that will successively print out the digits of the number, one after the other. Of course, in this case the machine will generally run on forever. To see if these definitions of computability are mathematically interesting, the first thing one would like to know is if there are any numbers that are not computable. We will now consider the limitations on our ability to actually compute numbers using such a device.

It can be shown that for a two-symbol Turing machine with $n$ possible states of the reading head, there are exactly $(4n + 4)^{2n}$ distinct programs that can be written, which, of course, coincides with the number of distinct Turing machines. This means that an $n$-state machine can compute at most this many numbers. Letting $n$ take on the values $n = 1, 2, 3, \ldots$, one may conclude that Turing machines can calculate at most a *countable* set of numbers, that is, a set whose elements can be put into a one-to-one correspondence with a set of positive integers (the "counting" numbers). But there are uncountably many real numbers; hence, we come to the perhaps surprising result that the vast majority of real numbers are not computable.

This counting argument is one way to show the existence of uncomputable numbers, albeit a somewhat indirect one. Turing himself used a more direct procedure based upon what's known as *Cantor's diagonal argument*. It goes like this. Consider the following listing of names: Smith, Otway, Arquette, Bethel, Bellman, and Imhoff. Now take the first letter of the first name and advance it alphabetically by one position. This gives a $T$. Then do the same for the second letter of the second name, the third letter of the third name and so on. The result is "Turing." Obviously, the name Turing could not have been on the original list, since it differs from each entry on that list by at least one letter.

Turing's argument for the existence of uncomputable numbers follows the same line of reasoning. Suppose you list all computable numbers between 0 and 1, say, written out by their decimal expansions. This can be done, since by the argument given a moment ago the computable numbers form a countable set, that is, they can be placed in one-to-one correspondence with the positive integers. Of course, such a list will be infinitely long, but here is the first part of it:

$$n_1 = 0.a_1a_2a_3a_4 \ldots ,$$
$$n_2 = 0.b_1b_2b_3b_4 \ldots ,$$
$$n_3 = 0.c_1c_2c_3c_4 \ldots ,$$
$$\vdots$$

where each of the quantities $a_1$, $b_i$, $c_i$ are integers between 0 and 9. Now advance the first digit of the first number, the second digit of the second number, and, in general, the $k$th digit of the $k$th number, each by one digit. In this way, a new number $N = 0.(a_1 + 1)(b_2 + 1)(c_3 + 1)\ldots$ is created. This number cannot have been on the original list, since it differs in at least one position from every number on that list. But, by definition, the list contains all computable numbers. Therefore, the new number must be uncomputable.

From the foregoing arguments we see that computable numbers are the exception and uncomputable numbers are the rule, since the computable numbers between 0 and 1, being a countable set, form a very "thin" subset of the uncountable set of real numbers constituting the interval between 0 and 1. This surprising fact shows that all the numbers we deal with in our everyday personal and professional lives, which by the very fact that we can write them down (that is, have a rule for generating their digits) means that they must be computable, form but a microscopically small subset of the set of all possible numbers. The overwhelming majority of numbers lie in a realm that's impossible to reach by following the rules of any type of computing machine. Now let's look at a couple of examples of specific uncomputable quantities.

### The Busy Beaver Game

Suppose you're given an input tape filled entirely with 0s. The challenge is to write a program for an $n$-state Turing machine such that: (1) the program must eventually halt, and (2) the program should print as many 1s as possible on the tape before it stops. Obviously, the number of 1s that can be printed is a function only of $n$, the number of states available to the machine. Equally clear is the fact that if $n = 1$, the maximum number of 1s that can be printed is only one, a result that follows immediately from the requirement that the program cannot run on forever. If $n = 2$, it is known that the maximum number of 1s that can be printed before the machine halts is 4, a result that again follows from the fact that the machine must eventually halt, while a 3-state machine can print 6 1s before stopping. Programs that print a maximal number of 1s before halting are called *n-state Busy Beavers*.

Table 4.2 gives the program for a 3-state Busy Beaver, while Figure 4.3 shows how this program can print six 1s on the tape before stop-

**Table 4.2** A 3-state Busy Beaver.

| State | Symbol Read | |
|:---:|:---:|:---:|
| | **0** | **1** |
| A | 1, R, B | 1, L, C |
| B | 1, L, A | 1, R, B |
| C | 1, L, B | 1, STOP |

ping. The reader should interpret the entries in Table 4.2 in the following way: Suppose the scanning head is in state A and reads the symbol 1 from the tape. According to the program of the Table 4.2, the machine should now take the action (1, L, C). This is Turing machine shorthand for saying that the scanning head should "print 1 on the square, move one square to the L(eft), and set the pointer to position C, that is, enter the internal state C." The other instructions in the table are interpreted similarly. The reader should also note that the position of the tape scanning head is shown in boldface in the figure.

Now for our uncomputable quantity. Define $BB(n)$ to be the number of 1s written by an $n$-state Busy Beaver program. Thus, the value of the Busy Beaver function $BB(n)$ is the greatest number of 1s that any halting program can write on the tape of an $n$-state Turing machine. We have already seen that $BB(1) = 1$, $BB(2) = 4$, and $BB(3) = 6$. From these results for small values of $n$, it looks as if the function $BB(n)$ might not have any particularly interesting properties as $n$ gets larger. But you can't judge a function just by its behavior for a few values of its argument, and detailed investigation has shown that

$$BB(12) \geq 6 \times 4096^{4096^{4096^{\cdots 4096^4}}}$$

where the number 4096 appears 166 times in the dotted region! So in trying to calculate the value of the Busy Beaver function for a 12-state Turing machine, we quickly arrive at a number so huge that it's effectively infinite. It turns out that for large enough values of $n$, the value of the Busy Beaver function for such values of $n$ exceeds the value of *any* computable function evaluated at those same values. In other words, the Busy Beaver function $BB(n)$ is an uncomputable function. So for a concrete example of an effectively uncomputable number, just

State

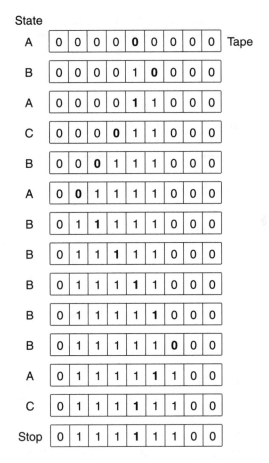

**Figure 4.3** The action of a 3-state Busy Beaver.

take a Turing machine with a large number of states $n$. Then ask for the value of the Busy Beaver function for that value of $n$. The answer is to all intents and purposes an uncomputable number.

Now to firmly fix the difference between the fact that something exists and our being able to explicitly display, that is, compute, that same something, let's consider another kind of game. This game will show that it's possible to prove the existence of a winning strategy for one of the players, while at the same time showing that the "winning" player will never be able to make effective use of this fact because he or she will not be able to *compute* the actions called for by the winning strategy at each stage of play.

## The Turing Machine Game

Assume there are two players, called rather unimaginatively A and B. These players take turns choosing positive integers as follows:

Step 1. Player A chooses a number $n$.
Step 2. Knowing $n$, Player B picks a number $m$.
Step 3. Knowing $m$, Player A selects a number $k$.

The rules then say that Player A wins the game if there is some $n$-state Turing machine that halts in exactly $m + k$ steps when started on a tape containing all 0s. Otherwise, Player B wins. It's fairly clear that this is a game of finite duration, since once the players have chosen their integers, all we need do is construct the $(4n+4)^{2n}$ Turing machines having $n$ states, running each of them for exactly $m + k$ steps to determine a winner. If even one of these Turing machine programs stops at step $m + k$, then Player A wins the game; if not, victory goes to Player B.

It's a fact from game theory that any zero-sum game of fixed, finite duration is determined, in the sense that there is a winning strategy for one of the players. In this case, it's Player B. Nevertheless, the Turing Machine Game is nontrivial to play since once the players have chosen their respective numbers, it is far from immediately evident which of them will be the eventual winner. This is because neither player has a definite rule, or an *algorithm* (that is, a computable strategy) that he or she can follow for winning the game. The proof of this fact relies on showing that *any* winning strategy involves computing the values of a function that grow faster than the values of the Busy Beaver function. But we already know that $BB(n)$ is uncomputable; hence, this new function must also be uncomputable. The reader is referred to the bibliography for brief descriptions of material containing the details of this proof.

A crucial aspect of both our definition of a computable number, as well as the Busy Beaver and Turing Machine Games, is that the programs involved have to stop after a finite number of steps. It's evident, I think, that you haven't really computed anything until the computational process terminates (even in the case of real numbers, where any finite computation generally yields only an approximation to the number you're trying to compute). It would be a pretty sad business, for example, if the accounting office's payroll program failed to stop so that your paycheck never got printed and mailed (although considerably less sad if the same fate befell the mortgage company's program when it came time to prepare your monthly statement). This simple observation leads to a key question in the theory of computation: Is there a general procedure (that is, an algorithm) that will tell us *in advance* whether or not a particular program will halt after a finite number of steps? This is the famous *Halting Problem.*

## Ad Infinitum?

Let's restate the Halting Problem in slightly more explicit terms: Given any Turing machine program $P$ and a set of input data $I$, is there a single program that accepts $P$ and $I$ as its input data, and that after a finite number of steps halts, at which point the final tape configuration then specifies whether or not $P$ will stop after a finite number of steps when processing the data $I$? Note carefully that what we're asking for here is a *single* program that will work for *all* instances of programs $P$ and *all* possible input data $I$. This is a kind of "metacomputational" question, since it asks about the existence of a program that will tell us about the characteristics of *all* programs.

To see that the question is far from trivial, suppose we have a program $P$ that reads a Turing machine tape and stops when it comes to the first 1. Thus, in essence the program says, "Keep reading until you come to a 1, then stop." In this case, the input data $I$ consisting entirely of 1s would result in the program stopping after the first step. On the other hand, if the input data was all 0s the program would never stop. Of course, in this situation we have a clear-cut procedure for deciding whether or not the program will halt when processing some input tape: just look at the tape. The program will stop if and only if the tape contains even a single 1; otherwise, the program will run on forever. This then is an example of a halting rule that works for any data set processed by this especially primitive program.

Unfortunately, most real computer programs are vastly more complicated than this, and it's far from clear by simple inspection of the program or the input tape what kinds of quantities will be computed as the program goes about its business. After all, if we knew beforehand what the program was going to compute at each step, we wouldn't have to run the program. Moreover, the stopping rule for real programs is almost always an implicit rule, saying something like, "If such and such a quantity satisfying this or that condition appears, stop; otherwise, keep computing." The essence of the Halting Problem is to ask if there exists any effective procedure that can be applied to the program and its input data to tell in advance whether or not the program's stopping condition will ever be satisfied. In 1936, Turing settled the matter once and for all with the Halting Theorem.

---

**HALTING THEOREM** *Given an arbitrary Turing machine program $P$ and an arbitrary set of input data set $I$, there does not exist a single Turing machine program that halts after a finite number of steps, and that will tell us if $P$ will ever finish processing the input $I$.* ■

### Turing–Church Thesis

The notion of a Turing machine finally put the idea of a computation on a solid mathematical footing, enabling us to pass from the vague, intuitive idea of an effective process to the precise, mathematically well-defined notion of an algorithm. In fact, Turing's work, along with that of the American logician Alonzo Church, forms the basis for what has come to be called the *Turing–Church Thesis*.

The content of the Turing–Church Thesis resides in its assertion that any quantity that can be computed can be computed by a suitable Turing machine. This claim is called a thesis and not a theorem because it's not really susceptible to proof. Rather, it's more in the nature of a definition, or a proposal, suggesting that we agree to equate our informal idea of carrying out a computation with the formal mathematical idea of a Turing machine. Stated more formally, we have the Turing–Church Thesis.

---

**TURING–CHURCH THESIS** *Every effective process is implementable by running a suitable program on a UTM.* ■

To bring home more forcefully this point of equating the carrying out of a computation with the operation of a Turing machine, it's helpful to draw an analogy between a Turing machine and a typewriter. (Is there anyone left who still uses such things?) Like a Turing machine, a typewriter is also a primitive device, allowing us to print sequences of symbols on a piece of paper that is potentially infinite in extent. A typewriter also has only a finite number of states that it can be in: uppercase and lowercase letters, red or black ribbon, different symbol balls, and so on. Yet despite these limitations, any typewriter can be used to type *The Canterbury Tales, Alice in Wonderland,* or any other string of symbols. Of course, it might take a Chaucer or a Lewis Carroll to be the first to tell the machine what to do, but it can be done. By way of analogy, it might take a very skilled programmer to tell the Turing machine how to solve difficult computational problems. However, the Turing–Church Thesis states, the basic model—the Turing machine—suffices for every type of question that is at all answerable by carrying out a computation.

It has probably not escaped the reader's attention that there is a quite evident parallel between the actions taken by a Turing machine as it goes about performing a computation and the steps a mathematician follows in proving a theorem using a chain of logical inferences. So let's now shift our attention for a while from the computational to the purely logical, our goal being to show that they are actually the same thing.

6

## Form and Content

During the 1928 International Congress of Mathematicians (ICM) in Bologna, Italy, the famed German mathematician David Hilbert threw down a challenge that would ultimately change forever the way we think about the relationship between what is logically provable and what is actually true.

At stake in Hilbert's 1928 address was the issue of whether or not it is possible to prove every true mathematical statement. What Hilbert was looking for was a kind of Truth Machine capable of settling every possible mathematical statement. Just feed the statement in at one end, turn the crank, and out the other end pops the answer: TRUE or FALSE. Ideally, in this setup the original statement would be either a true mathematical fact and, hence, logically deducible from the given assumptions and thus a theorem, or it would be false and, consequently, not a theorem, that is, its negation would be a theorem. In short, Hilbert's Truth

Machine would give a complete account of every possible mathematical assertion. In his Bologna talk Hilbert laid down the requirements for such a Truth Machine, or what's more pedantically termed an *axiomatic*, or *formal, logical system* along with the conviction that his research "Program" would ultimately yield a complete axiomatization of all mathematics.

With this challenge to the mathematical world, Hilbert was reemphasizing a different aspect of another problem he had posed at an earlier ICM gathering in Paris in 1900. With the conviction that unsolved problems are the lifeblood of any field of intellectual activity—and to mark the turn of the century—Hilbert listed 23 problems whose resolution he felt was of central importance for the development of mathematics. The second problem on this list involved proving that mathematical reasoning is reliable. In other words, by following the rules of mathematical reasoning, one should not be able to arrive at mutually contradictory statements; a proposition and its negation should not both be theorems. Of course, this self-consistency requirement is a necessary condition for any axiomatic system of the sort Hilbert had in mind to lead to logically coherent results, since Aristotle had shown long ago that if the system is inconsistent *any* assertion can be proved true or false as we please. And this is hardly a secure basis for any kind of reliable knowledge, mathematical or otherwise.

Less than 3 years after Hilbert's Bologna address, the young Austrian logician Kurt Gödel astonished the mathematical world by publishing a revolutionary paper turning Hilbert's fondest dream into his wildest nightmare. In his 1931 paper, Gödel showed that there exist true but unprovable mathematical statements. Put more prosaically, there is an eternally unbridgeable gap between what can be *proved* and what's *true*. The idea of establishing an axiomatic framework for all of mathematics— put forth as a primary goal of mathematics by Hilbert—was the starting point for Gödel's assault on proof. So we begin our story with a bit of background on Hilbert's Program for axiomatizing mathematical truth.

Hilbert believed that the way to eliminate the possibility of paradoxes like "This sentence is false" (that is, inconsistencies) arising in mathematics was to create an essentially "meaningless," purely syntactic framework within which to speak about the truth or falsity of mathematical statements. This would be a framework in which the mathematical statements are all expressed using purely abstract symbols having no intrinsic meaning other than what's given to them by definition in the framework itself. Such a framework is now termed a *formal system,* and it constitutes the historical jumping-off point for investigations of the gap between what can be proved and what is actually true in the universe of mathematics.

The "meaningless statements" of a formal system are composed of finite sequences of abstract *symbols.* The symbols are often termed the *alphabet* of the system, while the "words" of the system are usually called *symbol strings.* The symbols might be objects like ★, ✳, and ✠, or they might even be signs like 0 and 1. But in the latter case, it's absolutely essential to recognize that we're not talking yet about the *numbers* 0 and 1, but simply about the *numerals* 0 and 1. It's only when these symbols are given meaning as numbers that they acquire the properties we usually associate with the numbers 0 and 1. We'll come back to this point with a vengeance shortly. In a formal system, a finite number of these symbol strings are taken as the *axioms* of the system. To round things out, the system also has a finite number of *transformation rules,* or what are often called *rules of logical inference.* These rules specify how a given string of symbols can be converted into a new string of symbols.

The general idea of proof within a formal system is to start from one of the axioms and apply a finite sequence of transformations, thereby converting the axiom into a succession of new strings, where each string is either one of the given axioms of the system or is derived from its predecessors by application of the transformation rules. The last string in such a sequence is called a *theorem* of the system. The totality of all theorems constitutes what can be proved within the system. But note carefully that these so-called statements don't actually say anything; they are just strings of abstract symbols. We'll get to how the theorems acquire meaning in a moment. But first let's see how this setup works with a simple example.

Suppose the symbols of our system are the three objects ★ (star), ✠ (Maltese cross), and ✳ (sunburst). Let the two-element string ✠ ✳ be the sole axiom of the system. Letting $x$ denote an arbitrary finite string of stars, crosses, and sunbursts, we take the transformation rules of our system to be:

| Rule I: | $x$ ✳ | $\longrightarrow$ | $x$ ✳ ★ |
|---|---|---|---|
| Rule II: | ✠ $x$ | $\longrightarrow$ | ✠ $xx$ |
| Rule III: | ✳ ✳ ✳ | $\longrightarrow$ | ★ |
| Rule IV: | $x$ ★ ★ $x$ | $\longrightarrow$ | $xx$ |

In these rules, $\longrightarrow$ means "is replaced by." So, for instance, Rule I says that we can form a new string by appending a star to any string that ends in a sunburst. The interpretation of Rule IV is that any time two stars appear together in a string, they can be dropped to form a new string. Now let's see how these rules can be used to prove a theorem.

Starting with the single axiom ✠ ✳, we can deduce that the string ✠ ★ ✳ is a theorem by applying the transformation rules in the following order:

$\longrightarrow$ ✠✳ $\longrightarrow$ ✠✳✳ $\longrightarrow$ ✠✳✳✳✳ $\longrightarrow$ ✠★✳

(Axiom)  (Rule II)  (Rule II)  (Rule III)

Such a sequence of steps, starting from an axiom and ending at a statement like ✠ ★ ✳, is termed a *proof sequence* for the theorem represented by the last string in the sequence. Observe that when applying Rule III at the final step, we could have replaced the last three ✳s from the preceding string rather than the first three, thereby ending up with the theorem ✠ ✳ ★ instead of ✠ ★ ✳. The perceptive reader will have also noted that all the intermediate strings obtained in moving from the axiom to the theorem begin with ✠. It's fairly evident from the axiom and the action of the transformation rules for this system that every string will have this property. This is a *metamathematical* property of the system, since it's a statement *about* the system rather than one made *in* the system itself. The distinction between what the system can say from the inside (its strings) and what we can say about the system from the outside (properties of the strings) is of the utmost importance for Gödel's results.

Upon comparing the workings of a Turing machine program and the operations we just went through using the transformation rules of a formal system, one might say that there's no essential difference between the two. And so it is. The matchups showing what amounts to a perfect correspondence between Turing machines and formal systems are shown in Table 4.3.

We spoke earlier of Hilbert's famous *Entscheidungsproblem,* or Decision Problem, which asked if there is any algorithmic procedure for

**Table 4.3**  Turing machine-formal system correspondence.

| Turing Machine | Formal System |
| --- | --- |
| Tape symbols | Alphabet |
| Tape pattern | Symbol string |
| Input data | Axioms |
| Program instruction | Rule of inference |
| Output | Theorem |

deciding if a given symbol string is or is not a theorem of a particular formal system. Using the "isomorphism" in Table 4.3 between Turing machines and formal systems, Turing was able to translate the Decision Problem involving theorems in a formal system into its equivalent expression in the language of machines. We have already seen that this computing equivalent is the Halting Problem, whose negative solution implies the same sad answer to the Decision Problem. Since about now the right question to be asking yourself is: What does all this meaningless symbol manipulation have to do with everyday reality?, let's quickly turn our attention from matters of form to those of content.

The answer to how we get from form to content can be given in one word: *interpretation*. Let's focus our interest right now on the slice of everyday reality consisting of mathematical facts. Depending on the kind of mathematical structure under consideration (for example, euclidean geometry, elementary arithmetic, calculus, topology, and so forth), we have to make up a dictionary by which we can match up (that is, interpret) the objects constituting that mathematical structure, things like points, lines, and numbers, with the abstract symbols, strings, and rules of the formal system that we want to employ to represent the structure. By this dictionary construction step, we attach meaning, or semantic content, to the abstract, purely syntactic strings formed from the symbols of the formal system. Thereafter, all the theorems of the formal system can be interpreted as true statements about the associated mathematical objects. Figure 4.4 illustrates this crucial distinction between the purely syntactic world of formal systems and the meaningful world of mathematics.

Once this dictionary has been written and the associated interpretation established, then we can hope along with Hilbert that there will be a perfect, one-to-one correspondence between the true facts of the

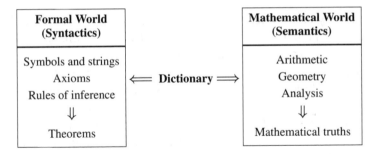

**Figure 4.4**  The formal and the mathematical worlds.

mathematical structure and the theorems of the formal system. Loosely speaking, Hilbert's dream was to find a formal system in which every mathematical truth translates into a theorem, and conversely. Such a system is termed *complete*. Moreover, if a mathematical structure is to avoid contradiction, a mathematical truth and its negation should never both translate to theorems, that is, be provable in the formal system. Such a system in which no contradictory statements can be proved is termed *consistent*. With these preliminaries in hand, we can finally describe Gödel's wreckage of Hilbert's Program.

## The Undecidable

By the time of Hilbert's 1928 Bologna lecture, mathematicians had already established that geometrical propositions as well as all other types of mathematical assertions could be recast as assertions about numbers. Thus, the problem of the consistency of mathematics as a whole was reducible to the determination of the consistency of arithmetic. That is, to the properties and relations among the natural numbers (the positive integers $1, 2, 3, \ldots$). So the problem became to give a "theory of arithmetic," that is, a formal system that was (1) finitely describable, (2) consistent, (3) complete, and (4) sufficiently powerful to represent all the statements that can be made about the natural numbers. By the term *finitely describable* what Hilbert meant was not only that the number and length of the axioms and rules of the system should be constructible in a finite number of steps, but also that every provable statement in the system—every theorem—should also be provable in a finite number of steps. This condition seems reasonable enough, since you don't really have a theory at all unless you can tell other people about it. And you certainly can't tell them about it if there are an infinite number of axioms, rules, and/or steps in a proof sequence.

A central question that arises in connection with any such formalization of arithmetic is to ask if there is a finite procedure by which we can decide the truth or falsity of every arithmetical statement. Thus, for example, if we make the statement "The sum of two odd numbers is always an even number," we want a finite procedure—essentially a computer program—that halts after a finite number of steps, telling us whether that statement is provable or not in some formal system powerful enough to encompass ordinary arithmetic. For example, in the ✠-★-✻ formal system considered above, such a decision procedure is given by the not-entirely-obvious conditions: "A string is a theorem if and only (1) if it begins with a ✠, (2) the remainder of the string consists solely of ★s and ✻s, and (3) the number of ✻s is not a multiple of 3."

Hilbert was convinced that a formalization of arithmetic satisfying the foregoing desiderata was possible, and his Bologna manifesto challenged the international mathematical community to find or create it. But in 1931, less than 3 years after Hilbert's Bolognese call to arms, Kurt Gödel published the following metamathematical fact, perhaps the most famous mathematical (and philosophical) result of this century:

---

**GÖDEL'S THEOREM—INFORMAL VERSION**   *Arithmetic is not completely formalizable.*   ∎

Remember that for arithmetic there are an infinite number of ways we can choose a finite set of axioms and rules of inference in a formal system so as to attempt to mirror syntactically the mathematical truths about numbers. Gödel's result says that *none* of these choices will work; there does not and cannot exist a formal system satisfying all the requirements of Hilbert's Program. In short, there are no rules for generating *all* the truths about the natural numbers.

In arriving at his proof of the incompleteness of arithmetic, Gödel's first crucial observation was to recognize the importance of Hilbert's insight that every formalization of a branch of mathematics is itself a mathematical object in its own right, since what we mean when we say we have "formalized" something is that we have created a mathematical framework within which to speak about whatever it is we wanted to formalize. So if we create a formal system intended to capture the truths of arithmetic, that formal system can be studied not just as a set of mindless rules for manipulating symbols, but also as an object possessing mathematical, that is, semantic, as well as syntactic properties. In particular, since Gödel was interested in the relationships between numbers, his goal was to represent any formal system purporting to encompass arithmetic within arithmetic itself. Basically, this involved showing how to code *any* statement about numbers and their relationships by a unique number itself. Thus, what Gödel saw was a way to mirror all statements about relationships between the natural numbers by using these very same numbers themselves.

This mirroring idea is probably more familiar in the context of ordinary language where we use words *in* the English language to speak *about* language. For example, we use words to describe properties of words like whether they are nouns or verbs, and we discuss the structure of, say, a treatise on English grammar, which consists of words, by employing other words of the English language. Thus, in both cases we are making use of language in two different ways: (1) as a collection of *uninterpreted* strings of alphabetic symbols that are manipulated according to the rules of English grammar and syntax, and (2) as a set of *interpreted* strings having a meaning within the context under discussion. So the key notion is that the very same objects can be considered in two quite distinct ways, opening up the possibility for that object to speak about itself. In passing, let me note that the very same dual-level idea pertains to the symbols and their interpretations in the genetic material (the DNA) of every living cell. The nucleotide bases A, G, C, and T on the DNA strand can either be interpreted as instructions for building the proteins from which every living organism is formed, or they can simply be copied without interpretation as, for example, when the DNA is replicated in the process of cell division. Gödel discovered how to do this same trick with mirrors using the natural numbers.

To see how Gödel's method works, let's consider a somewhat streamlined version of the language of symbolic logic as found in the monumental treatise *Principia Mathematica* by Bertrand Russell and Alfred North Whitehead. This slimmed-down version is due to Ernest Nagel and James R. Newman. In this toy version of the language of logic there are elementary signs and variables. To follow Gödel's scheme, suppose there are the 10 logical signs shown in Table 4.4, each with its Gödel code number, an integer between 1 and 10.

In addition to the elementary signs, the language of the *Principia* contains logical variables that are linked through the signs. These variables come in three different flavors, representing a kind of hierarchical ordering that depends upon the exact role the variable plays in the overall logical expression. Some variables are *numerical,* meaning that they can take on numerical values. For other variables we can substitute entire logical expressions or formulas (*sentential* variables). Finally, we have what are called *predicate* variables, which express properties of numbers or numerical expressions like *prime, odd,* or *less than.* All the logical expressions and provability relations in *Principia Mathematica* can be written using combinations of these three types of variables, connecting them via the logical signs. For our streamlined version of *Principia,* there are only 10 logical signs, although in the real case there are quite a few more. In this scaled-down version of *Principia Mathematica,* Gödel's numbering system would code numerical variables by prime numbers greater than 10, sentential variables by the squares of primes greater than 10, and predicate variables by the cubes of primes greater than 10.

To see how this numbering process works, consider the logical formula $(\exists x)(x = \mathbf{s}y)$, which translated into plain English reads: "There exists a number $x$ that is the immediate successor of the

Table 4.4   Gödel numbering of the elementary logical signs.

| Sign | Gödel Number | Meaning |
|------|--------------|---------|
| ~ | 1 | Not |
| ∨ | 2 | Or |
| ⊃ | 3 | If . . . then |
| ∃ | 4 | There exists |
| = | 5 | Equals |
| 0 | 6 | Zero |
| s | 7 | The immediate successor of |
| ( | 8 | Punctuation |
| ) | 9 | Punctuation |
| / | 10 | Punctuation |

number $y$." Since $x$ and $y$ are numerical variables, the Gödel coding rules dictate that we make the assignment $x \rightarrow 11$, $y \rightarrow 13$, since 11 and 13 are the first two prime numbers larger than 10. The other symbols in the formula can be coded by substituting numbers using the correspondences in Table 4.4. Carrying out this coding yields the sequence of numbers $\{8, 4, 11, 9, 8, 11, 5, 7, 13, 9\}$, formed by reading the logical expression symbol by symbol and substituting the appropriate number according to the coding rule. This sequence of 10 numbers pins down the logical formula uniquely. But since number theory—that is, arithmetic—is about the properties of single numbers, not sequences of numbers, one would like to be able to represent the formula in an unambiguous way by a single number. Gödel's procedure for doing this is to take the first 10 prime numbers (since there are 10 symbols in the formula) and multiply them together, each prime number being raised to a power equal to the Gödel number of the corresponding element in the formula. Since the first 10 prime numbers in order are 2, 3, 5, 7, 11, 13, 17, 19, 23, and 29, we make the substitutions ($\rightarrow 2^8$, $\exists \rightarrow 3^4$, $x \rightarrow 5^{11}$, and so on. The final Gödel number for the above formula is then

$$(\exists x)(x = \mathbf{s}y) \rightarrow 2^8 \times 3^4 \times 5^{11} \times 7^9 \times 11^8 \times 13^{11} \times 17^5 \times 19^7 \times 23^{13} \times 29^9.$$

Using this kind of numbering scheme, Gödel was able to attach a unique number to each and every statement and sequence of statements about arithmetic that could be expressed in the logical language of *Principia Mathematica*. So with this scheme, every possible proposition about the natural numbers can itself be expressed as a number, thereby opening up the possibility of using arithmetic to examine its own truths.

Deep insight and profound results necessarily involve seeing the connection linking several ideas at once. In the proof of Gödel's Theorem there are two crucial notions that Gödel had to juggle simultaneously, Gödel numbering being the first. Now for the second Big Idea, the substitution of the notion of a mathematical proof for the everyday concept of truth, together with the translation of a verbally stated logical paradox into an arithmetic statement.

Logical paradoxes of the sort that worried Hilbert are all based on the notion of self-reference, that is, they all involve statements that refer to themselves. The granddaddy of all such conundrums is the so-called Liar Paradox, one version of which is

| This sentence is false. |

Now why is this a paradox? Well, in everyday parlance for something to be "false" means that it does not correspond to reality. The sentence says that it is false. If that assertion does not correspond to reality, then the sentence must be true. On the other hand, if the sentence is true, this means that what it says does correspond to reality. But this true sentence says that it is false. Therefore, the sentence must indeed be false. So whether you assume that the sentence is true or that it's false, you are forced into concluding the opposite! Thus, the Paradox of the Liar.

What Gödel wanted to do was find a way to express such paradoxical self-referential statements within the framework of arithmetic. He needed such a statement in order to display an exception to Hilbert's thesis that all true assertions should be provable in a formal system. However, a statement like the Liar Paradox involves the notion of truth, something that logician Alfred Tarski had shown earlier could not be captured within the confines of a formal system. Enter Gödel's Big Idea #2.

Instead of dealing with the eternally slippery notion of truth, Gödel had the insight to think of replacing "truth" by something that is formalizable: the notion of *provability*. Thus, he modified the Liar Paradox above into the Gödel sentence

> **This statement is not provable.**

This sentence, of course, is a self-referential claim about a particular "statement," the statement mentioned in the sentence. However, by his numbering scheme Gödel was able to code this assertion by a corresponding self-referential, metamathematical statement expressed in the language of arithmetic itself. Let's follow through the logical consequences of this mirroring.

If it turns out that the statement referred to *is* provable, then by Gödel's equating of truth with proof the statement must be true. Therefore, what it says must be true. But what it says is that it is *not* provable. Consequently, the statement and its negation are both provable, implying an inconsistency in our logical scheme of proof. On the other hand, if the statement referred to is not provable, then what it asserts is indeed the case, that is, the statement is true, but unprovable. Thus, there is a true statement that is not provable, implying that the formal system we are using for proving statements is incomplete.

Remember what Gödel showed was how to translate this verbal self-referential statement into an equivalent statement within the formal system accepted by mathematicians for proving statements of arithmetic. This means that the logical conclusions we have just drawn about inconsistency and incompleteness apply to the entire mathematical apparatus of numbers. Thus, if the formal system used for arithmetic is consistent, then it must necessarily be incomplete.

Gödel was able to show that for *any* consistent formal system powerful enough to allow us to express all statements of ordinary arithmetic, such a Gödel sentence must exist; consequently, the formalization must be incomplete. The bottom line then turns out to be that in *every* consistent formal system powerful enough to express all relationships among the whole numbers, there exists a statement that cannot be proved using the rules of the system. Nevertheless, that statement represents a true assertion about numbers, one that we can see is true by jumping outside of the system. Almost as an aside, Gödel also showed how to construct an arithmetical statement $A$, which translates into the metamathematical claim "arithmetic is consistent." He then demonstrated that the statement $A$ is not provable, implying that the consistency of arithmetic cannot be established by using any formal system representing arithmetic itself. Putting all these different notions together, we finally arrive at the following theorem.

---

**GÖDEL'S THEOREM—FORMAL LOGIC VERSION** *For every consistent formalization of arithmetic, there exist arithmetic truths that are not provable within that formal system.* ∎

Since the steps leading up to Gödel's startling conclusions are both logically tricky and intricately intertwined, let me summarize the principal landmarks along the road in Table 4.5.

The proverbial perceptive reader will by now have noticed the striking similarity between the results of Turing and Gödel. But for those who haven't, let me spell out this parallel more explicitly. Here are restatements of both results that capture the distilled essence of the two theorems:

**Table 4.5**   The main steps in Gödel's proof.

---

**Gödel Numbering:** Development of a coding scheme to translate every logical formula and proof sequence in *Principia Mathematica* into a "mirror image" statement about the natural numbers.

**Liar Paradox:** Replace the notion of "truth" with that of "provability," thereby translating the Liar Paradox into the assertion "This statement is unprovable."

**Gödel Sentence:** Show that the sentence "This statement is unprovable" has an arithmetical counterpart, its Gödel sentence *G*, in every conceivable formalization of arithmetic.

**Incompleteness:** Prove that the Gödel sentence *G* must be true, but unprovable, if the formal system is consistent.

**No Escape Clause:** Prove that even if additional axioms are added to form a new system in which *G* is then provable, the new system with the additional axioms will have its own unprovable Gödel sentence.

**Consistency:** Construct an arithmetical statement *A* asserting that "arithmetic is consistent." Prove that this arithmetical statement is not provable, thus showing that arithmetic *as a formal system* is too weak to prove its own consistency.

---

**GÖDEL'S THEOREM**   *For any consistent formal system $\mathcal{F}$ purporting to settle, that is, prove or disprove, all statements of arithmetic, there exists an arithmetical proposition that can be neither proved nor disproved in this system. Therefore, the formal system $\mathcal{F}$ is incomplete.* ∎

---

**THE HALTING THEOREM**   *For any Turing machine program $\mathcal{H}$ purporting to settle the halting or nonhalting of all Turing machine programs, there exists a program $\mathcal{P}$ and input data $\mathcal{I}$ such that the program $\mathcal{H}$ cannot determine whether or not $\mathcal{P}$ will halt when processing the data $\mathcal{I}$.* ∎

When placed side by side in this fashion, it becomes fairly evident, I think, that the Halting Theorem is simply Gödel's Theorem expressed in terms of computing machines and programs instead of in the language of logical deductive systems.

Turing's solution of the Halting Problem and the equivalence of the Halting Problem to Hilbert's Decision Problem, together with the faithful correspondence between Turing machines and formal systems, allows us to conclude that there cannot exist a Turing machine program that will print out all the true statements of arithmetic.

Gödel's results show that there are statements about numbers that we can see to be true—yet which cannot be *proved* by following a chain of logical reasoning. Put another way, no single set of rules will ever "fence in" all possible true statements about numbers; truth is strictly bigger than proof. Some philosophers have taken this to mean that the power of the human mind somehow transcends the power of deductive reasoning. It's but a small step from here to conclude that we will never create a computing machine with powers equal to the human mind, since computing machines are completely equivalent in the truths they can generate to those that can be obtained by following the rules of a formal logical system. Let's take a harder look at this Gödelian argument against machines having mental states and thinking just like you and me.

## Manufactured Minds

In 1950, Alan Turing published the paper "Computing Machinery and Intelligence," which sparked off a debate that rages to this day over the question: Can a machine think? In addition to its pivotal role in drawing attention to the matter of machine intelligence, Turing's paper was notable for its introduction of an operational test for deciding whether or not a machine really was thinking—human style. This criterion, now termed the *Turing Test,* is unabashedly behavioristic in nature, involving the machine's fooling a human interrogator into thinking it is actually a human solely on the basis of what the machine *does,* ignoring completely the machine's material structure and, in particular, *how* it does what it does. Turing's rationale for proposing what he called the "Imitation Game" was that the only way we have for deciding whether or not other humans are thinking is to observe their behavior. If this criterion is good enough to decide if humans are thinking, then fairness to machines dictates that it's good enough for them, too.

On November 8, 1991, the Boston Computer Museum held the world's first hands-on Turing Test, in which eight programs conversed with human inquisitors on a restricted range of topics that included women's clothing, romantic relationships, and Burgundy wines. At the day's end, the judges awarded first prize to a program called *PC Therapist III,* which was designed to engage its questioner in a whimsical conversation about nothing in particular. To illustrate the kind of dialogue such programs generate, here is a snippet of interchange between another contestant, the program *Good Mood,* which is designed to serve as an electronic psychoanalyst, and one of the interrogators. *Good Mood* begins by saying to the interrogator: "Perhaps you're not getting enough affection from your partner in the relationship." The interrogator replies, "What are the key elements that are important in relationships in order to prevent conflict or problems?" "I think you don't think I think," responded the machine.

This kind of interchange did little to fool the judges, most of whom said they were able to spot the mistakes rooted in a lack of everyday common sense that immediately singled out the computer programs from the responses displayed on the terminals controlled by the humans. Nevertheless, the overall conclusion from this historic experiment was that perhaps the Turing Test isn't as difficult as many people originally thought, since even the primitive programs in this contest managed to fool some of the judges most of the time. Of course, we should keep in mind that this wasn't a *true* Turing Test, since the domains of discourse were severely restricted. But it was still a pretty good initial approximation.

The Turing Test represents a third-party perspective on human intelligence, involving one standing outside the system and discerning humanlike intelligence in a machine by observing only the machine's behavioral output. The test says nothing about the internal constitution of the machine, how its program is structured, the architecture of the processing unit, or its material composition. In Turing's view of intelligence, only externally observed behavior counts. And if you behave like us, then you are a thinking machine.

In 1989, theoretical physicist Roger Penrose published *The Emperor's New Mind,* a book whose central argument is that the human mind is capable of transcending rational thought, hence can never be duplicated in a machine. Before going on, let me note that we are using the term *rational thought* in the strong sense of following rules or an algorithm to arrive at a result by a process of logical deductive inference. There is no connection here with the everyday, economic interpretation of rationality as relating to self-interest or prudent action. So what Penrose argues is that at least some human thought processes do not involve following any kind of rule. He justified this claim by a wildly speculative appeal to quantum processes in the human brain as the basis of consciousness and intelligence. A key ingredient in Penrose's argument is Gödel's result showing that there are true statements of arithmetic that the human mind can know but that cannot be the end result of following a fixed set of rules (that is, a computer program). Here Penrose is reviving a line of attack originally put forward by Oxford philosopher John Lucas in 1961. To understand a little bit better the relationship between Gödel's Theorem and the possible limits to machines ever displaying humanlike intelligence, let's take a little harder look at the chain of reasoning underlying the claims of both Lucas and Penrose.

Gödel's theorems show that any reasonably rich formal system is incomplete and that the consistency of such a system cannot be proved within the system itself. Furthermore, in Turing's work we saw that formal systems and machines are equivalent in what they can do by way of producing logically valid conclusions from given assumptions. Ergo, computers are subject to the same limitations that Gödel imposed on any formal system. Lucas and Penrose then use this fact to jump to the conclusion that machines are inherently limited in what they can do and, in particular, there are statements that the mind knows to be true but that the machine cannot prove. Interestingly enough, Turing anticipated this kind of objection to machine intelligence in his classic 1950 paper on thinking machines, in which he replied that people may well be subject to similar limitations. But John Lucas wasn't convinced by Turing's response, and in his 1961 paper titled "Minds, Machines, and Gödel" he attempted to strengthen the Gödelian argument against the view that the mind is a machine.

The heart of both Lucas's and Penrose's arguments takes the following course. By standing outside the incomplete, consistent formal system, Gödel's results imply that humans can *know* that there exists some true, but unprovable, statement. But the machine cannot prove this fact; hence, a human can beat every machine since such a true, but unprovable, statement exists for every machine. Furthermore, if the human mind were nothing more than a formal system, by Gödel's second theorem the mind could not prove its own consistency. But humans do assert their own consistency. Consequently, the mind must be more than a machine.

As with virtually all philosophical debates, the arguments against Lucas hinge upon the precise meaning he gives to terms like *machine, consistency,* and *mind,* as well as the hidden assumptions underpinning his conclusions. For example, Paul Benacerraf points out that Lucas has too limited a view of machines, since any machine that could reprogram itself would be exempt from the Gödel argument. Furthermore, it is also noted that Lucas *assumes* that mind is consistent. In fact, this is far from obvious, as the following paradox constructed by C. H. Whitley shows.

Consider the sentence "Lucas cannot consistently assert this sentence." Lucas cannot assert the truth of this sentence even though he can clearly see that it's true. Why? Because if Lucas could assert it, then that fact would undermine his assumed consistency. Thus, either there is something that Lucas can see to be true but can't assert, or he is inconsistent. Whitley concludes that Lucas holds too high a regard for humans since even if there is an unprovable statement that a specific machine cannot assert, humans can't always do it either.

Other arguments countering Lucas claim that he errs in his application of Gödel's results. For instance, Gödel's Theorem proves only that a machine $M$ cannot prove the Gödel sentence of $M$ from *its* axioms and according to *its* rules of inference. But the human mind cannot prove the Gödel sentence either, at least not by using the axioms and rules of inference available to the machine. Furthermore, Lucas doesn't show that he can find a flaw in any machine, but only in any machine that the mechanist can construct. So in concluding this discussion of minds, machines, and artificial intelligence, it's of some interest to listen to what Gödel himself had to say about the matter.

Unfortunately, Gödel was rather reclusive and secretive, especially in his later years, and his only published statement on the topic comes from a lecture delivered to the American Mathematical Society in 1951:

> The human mind is incapable of formulating (or mechanizing) all its mathematical intuitions, i.e., if it has succeeded in formulating some of them, this very fact yields new intuitive knowledge, e.g., the consistency of this formalism. This fact may be called the "incompletability" of mathematics. On the other hand, on the basis of what has been proved

so far, it remains possible that there may exist (and even be empirically discoverable) a theorem-proving machine which in fact *is* equivalent to mathematical intuition, but cannot be *proved* to be so, nor even be proved to yield only *correct* theorems of finitary number theory.

Thus, Gödel leaves open the possibility of the existence of a theorem-proving machine, and even concedes that it may be possible to discover such a machine by empirical investigation. Thus, he says that there could exist a machine whose abilities equaled human mathematical intuition, but whose program we could never understand. Nonetheless, we would be able to set up conditions leading to the existence of such a machine, for example, by evolution. So, machines too complex to design could nevertheless exist.

With this statement about programs too complex for the human mind to design or understand, Gödel calls for us to face squarely the issue of how to characterize and measure the complexity of a computer program. This investigation has led to new and deep insights into Gödel's original results about the limitations of mechanical reasoning. These new results, in turn, suggest a variety of questions about the limits of science itself as a way of getting at the "scheme of things" in a real-world sense. So let's turn our attention to a consideration of how "complex" things can get before human reasoning comes up against its limits.

## Omega Is the End

Some years ago, *Scientific American* columnist Martin Gardner introduced a distinction between "dull" and "interesting" numbers. Interesting numbers, according to Gardner, are those that have some peculiar pattern or property that separates them from all other numbers. Dull numbers, on the other hand, are all those numbers that are not interesting. Gardner then went on to show the paradoxical nature of this dichotomy, by showing that dull numbers cannot possibly exist. His argument was to first list the integers in order, letting $D$ stand for the first dull number on the list. But the very fact that $D$ is the first dull number makes it interesting! Therefore, there can be no dull numbers.

One way to break out of this kind of paradoxical circle is to define a number to be interesting if it can be computed by a program of shorter length (that is, fewer bits) than the number itself. Such a short program would then encapsulate some special feature of the number through which the number could be distinguished from the general run of numbers. By way of contrast, dull numbers would be those that are algorithmically incompressible in the sense that they contain no pattern that could be exploited to reduce the size of their minimal program. It's reasonable to term such "patternless" numbers *random*. Clearly, most numbers are random in this sense, since for any $n$ there are more than twice as many numbers of length no greater than $n$ bits as there are numbers no longer than $n - 1$ bits to serve as shorter descriptions. This is because there are exactly $1 + 2 + 4 + \cdots + 2^n$ numbers less than or equal to $n$ bits in length. So the ratio of numbers no more than $n$ bits in length to those no longer than $n - 1$ bits is

$$\frac{1 + 2 + 4 + \cdots + 2^n}{1 + 2 + 4 + \cdots + 2^{n-1}} = 2 + \frac{1}{2^n - 1} > 2.$$

The idea of using the length of the shortest computer program to characterize the complexity of a number was introduced independently by Gregory Chaitin and Andrei Kolmogorov in the 1960s. Chaitin later used the idea to prove the following version of Gödel's Theorem:

---

**GÖDEL'S THEOREM—COMPLEXITY VERSION**    *Although almost all numbers are random, there is no formal axiomatic system that will allow us to prove this fact.* ∎

We can express this remarkable result more explicitly as follows. Suppose we have a formal system whose axioms and rules of inference require $n$ bits to describe, that is, the Turing machine program for this system is $n$ bits in length. Then this system cannot prove the randomness of any number longer than $n$ bits. Why? Well, suppose there was a proof in this system establishing the randomness of a number that is substantially longer than $n$ bits. Then we would have an $n$-bit program that could print out this random number. But, by definition, the randomness of the number means that there cannot exist a program shorter than the number in question that can produce that number. Thus, we have a contradiction, showing that such a program cannot exist after all. What this adds up to is a mathematical proof of the rather commonsense fact that you can't get more information out of a system than you put into it. As Georgia Tech physicist Joseph Ford once put it, "A 10-pound theory can no more generate a 20-pound theorem than a 100-pound pregnant woman can birth a 200-pound child." (Here, of course, Ford is referring to the weight of the woman while carrying her unborn child.)

## The Halting Probability

Earlier, we gave examples of uncomputable quantities in terms of the Busy Beaver Function and the Turing Machine Game. Chaitin has exploited the notion of algorithmic complexity in order to produce an even more dramatic example of such a number, something that he calls $\Omega$ ("omega"). It is closely related to the Halting Problem discussed above. Chaitin defines $\Omega$ to be the probability that a randomly generated program for a universal Turing machine will halt. Here when we say that the program is generated "randomly," it means that whenever the computer asks for another bit of input, we simply toss a fair coin and give the computer a 1, say, if the coin comes up heads and a 0 if the coin reads tails. Since Table 4.1 has shown us that every Turing machine program can be coded as a binary string, this procedure makes perfectly good sense. Now suppose that we generate a large number of such random programs, each statistically independent of the others, and run each of them for, say, 1 million steps. The ratio of those programs that stop before 1 million steps to those that don't then constitutes a lower bound on the number $\Omega$. If one lets the number of steps increase from 1 million to infinity, this ratio will then converge exactly to $\Omega$.

The quantity $\Omega$ is a perfectly well-defined number between 0 and 1, once the specific language for the UTM has been specified. For the types of programming languages we are familiar with, such as Fortran, C, or Pascal, $\Omega$ is likely to be close to 1, since a program generated at random in one of these languages is far likelier to crash immediately due to a syntax error than to go into an endless loop. Nevertheless, it can be shown that after the first few digits $\Omega$ would look very random indeed. What makes $\Omega$ interesting insofar as the theory of computation goes is that it encodes the Halting Problem is a very compact form. For example, knowing the first $n$ bits of $\Omega$ enables us to solve the halting problem for any program up to $n$ bits in length. Here's how.

Suppose we want to solve the Halting Problem for a particular randomly generated $n$-bit program $\mathcal{P}$. The program $\mathcal{P}$ corresponds to a specific sequence of $n$ coin tosses, having probability $2^{-n}$. If $\mathcal{P}$ halts, this much probability is then contributed to the total halting probability $\Omega$. Now let $\Omega_n$ be the known first $n$ bits of $\Omega$. This means that $\Omega$ is larger than $\Omega_n$ and smaller than $\Omega_n + 2^{-n}$. Thus, in order to decide whether or not the program $\mathcal{P}$ halts, we begin a systematic search for all programs that halt—regardless of their length. We do this by first running one program and then another for longer and longer times until we have seen enough halting programs to account for more than a fraction $\Omega_n$ of the total probability. Figure 4.5 illustrates the general idea. Then either $\mathcal{P}$ is among the programs that have halted so far or it will never halt. This is because its halting would drive up the total probability beyond its known upper bound of $\Omega_n + 2^{-n}$.

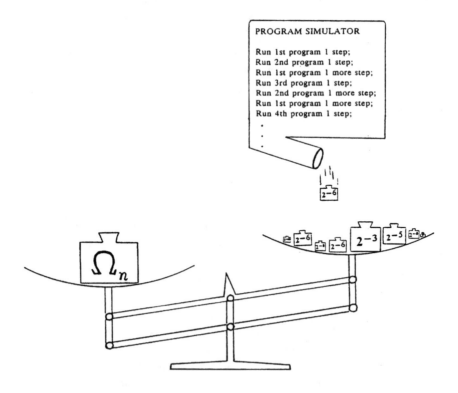

**Figure 4.5**  Using $\Omega$ to solve the Halting Problem.

Not only can knowledge of $\Omega$ be used to solve the Halting Problem, but it can also be used to settle many of the famous unproved conjectures in mathematics. For example, the famous Riemann Hypothesis, asserting that all the zeros of the Riemann zeta function lie on the line $\mathrm{Re}\, z = \frac{1}{2}$ in the complex plane, is equivalent to the assertion that some program, which searches systematically for zeros off the magical line, will never halt. Conjectures of this sort are usually simple enough that they can be encoded in the halting of *small* programs, generally not longer than a few thousand bits. Consequently, we could resolve all these conjectures if only we knew the first few thousand bits of $\Omega$.

This whole line of argument can be extended to classes of statements of the form: "Proposition $P$ is provable in formal system $\mathcal{F}$." If we suppose that proposition $P$ and system $\mathcal{F}$ together require $n$ bits to describe, then there is a certain program of about $n$ bits in length that will halt if and only if $P$ is provable in $\mathcal{F}$. Thus, for any proposition $P$ and system $\mathcal{F}$ simple enough to be comprehensible by the human mind, the first few thousand bits of $\Omega$ are sufficient to decide whether (1) $P$ is provable in $\mathcal{F}$, (2) not-$P$ is provable in $\mathcal{F}$, or (3) $P$ is independent of the axioms of $\mathcal{F}$.

At this point in our narrative, it will probably strike the reader as odd—or even paradoxical—that $\Omega$ can contain so much information about the Halting Problem and yet be computationally indistinguishable from the type of random sequence one gets by tossing a coin. The explanation for this oddity is as straightforward as it is remarkable. The fact is that $\Omega$ is an absolutely informative message, in the sense that all redundancy has been squeezed out of it. So it just appears random because it hasn't got a single bit of "fat" left in it. This means that $\Omega$ is pure information.

The fact that $\Omega$ contains the answer to almost any question that one can pose has moved IBM physicist Charles Bennett to call $\Omega$ a "cabalistic" number. It can be known of, but not known, through human reason. As Bennett remarks,

It embodies an enormous amount of wisdom in a very small space, inasmuch as its first few thousand digits ... contain the answers to more mathematical questions than could be written down in the entire universe, including all finitely-refutable conjectures. Its wisdom is useless precisely *because* it is universal; the only known way of extracting the solution to one halting problem from $\Omega$ . . . is by embarking on a vast computation that would at the same time yield solutions to all other equally simply-stated halting problems . . . . Ironically, although $\Omega$ cannot be computed, it might accidentally be generated by a random process, for example, a series of coin tosses, or an avalanche that left its digits spelled out in the pattern of boulders on a mountainside. The initial few digits of $\Omega$ are thus probably already recorded somewhere in the universe. Unfortunately, no mortal discoverer of this treasure could verify its authenticity or make practical use of it.

On this bittersweet note, let's now turn our attention from cabalistic quantities that can be known about but never computed to quantities that can be computed, in principle, but that will probably never be known. Thus far, we have focused on the distinction between quantities that are computable—given an infinite amount of time and storage—and those quantities like the Busy Beaver function or $\Omega$ that are logically uncomputable even with unlimited computational resources at our disposal. This distinction is of great theoretical interest; however, from a practical point of view it may well turn out that there are quantities that are theoretically computable by the Turing machine criterion, but that would take an length of time greater than the age of the universe for even the fastest of supercomputers to produce. We devote the remainder of the chapter to an account of this question of *computational intractability*.

## Tough Times

A famous problem of recreational mathematics is the so-called *Tower of Hanoi*. In this problem, there are three pegs $A$, $B$, and $C$, with $N$ rings of decreasing radii piled on the first peg $A$. The other two pegs are initially empty. The task is to transfer the rings from $A$ to $B$, perhaps using peg $C$ in the process. The rules stipulate that the rings are to be moved one at a time, and that a ring can never be placed upon one smaller than itself. Figure 4.6 shows the problem for the case of $N = 3$ rings.

In this case of three rings, it's not too hard to see that the sequence of seven moves

$$A \to B, \qquad A \to C, \qquad B \to C, \qquad A \to B,$$

$$C \to A, \qquad C \to B, \qquad A \to B$$

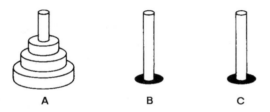

**Figure 4.6**  The Tower of Hanoi Game.

achieves the desired transfer of rings. And, in fact, it can be shown that there is a general algorithm, that is, a program, solving the game for any number of rings $n$. This program shows that the minimal number of transfers required is $2^n - 1$. Amusingly, the original version of this puzzle, dating back to ancient Tibet, involves $n = 64$ rings. So it's not hard to see why the Tibetan monks who originated the game claim that the world will end when all 64 rings are correctly piled on peg $B$. To carry out the required $2^{64} - 1$ steps, even performing one ring transfer every 10 seconds, would take well over *5 trillion* years! Thus, the number of steps needed for the solution of the Tower of Hanoi problem grows exponentially with the number of rings $n$. This is an example of a "hard" computational problem—one in which the number of computational steps needed to obtain a solution increases exponentially with the "size" of the problem.

By way of contrast, a computationally "easy" problem is the sorting of a deck of playing cards into the four suits in ascending order. First go through the deck until you find the ace of spades. Set it aside and then go through the remaining cards until you find the two of spades, which you also set aside. As one continues in this fashion, the deck is fully sorted. The worst that can happen with this sorting scheme is that the ace of spades is the last card in the unsorted deck, the two of spades is the next-to-last card, and so on. So starting with $n$ cards, you would have to examine at most $n^2$ cards. Thus, the number of steps needed to completely sort the deck is a quadratic function of the size of the problem, that is, the number of cards in the deck.

In the mid-1960s, J. Edmonds and A. Cobham introduced the idea of classifying the computational difficulty of a problem according to whether or not there exists an algorithm for solving the problem that requires at most a polynomial number of steps in the size of the problem. "Easy" problems can be solved in polynomial time; "hard" problems, on the other hand, require an exponentially increasing number of steps as the problem size grows. The difference in these two rates of growth is shown in Figure 4.7, where we see that a polynomial function may actually exceed an exponentially increasing one for small values of the size $n$. But as $n$ continues to grow, the exponential always wins out. To be definite about things, an algorithm is said to run in *polynomial time* if there are fixed integers $A$ and $k$ such that for a problem of size $n$, the computation will be completed in at most $An^k$ steps. For future reference, we let $P$ be the class of all problems that run in polynomial time. Algorithms that do not run in polynomial time are said to run in *exponential time*. Therefore an algorithm that requires $2^n$ or $n!$ steps to solve a problem of size $n$ is an exponential-time algorithm.

As an aside, it's worth noting here that this classification into hard and easy problems may be a bit misleading when it comes to hands-on computation. For example, an algorithm that has $A = 10^{50}$ and $k = 500$ will still run in polynomial time. But such an algorithm would hardly be "efficient" in any practical sense. On the other hand, an algorithm

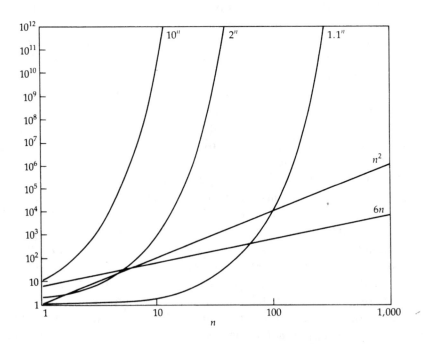

**Figure 4.7** Polynomial and exponential growth of functions.

for the same problem that ran in a number of steps on the order of $2^n$ might well be preferable to the polynomial-time procedure for small to moderate values of $n$—even though it is theoretically "inefficient." So the reader should keep in mind that what we are talking about here is the *theory* of computation, not its practice. However, in practice it seems to be the case that problems are solvable either only by exponential-time algorithms or by polynomial-time algorithms that run on the order of something like $10n^2$ or $50n^3$ steps—or less.

## P and NP

There is a very important class of problems known as $NP$, which stands for "nondeterministic polynomial time." Let me hasten to point out that this does *not* mean that there is something fuzzy, random, or indeterminate about such problems. Rather, a problem is classified as $NP$ if it's possible to verify a proposed solution of the problem in a number of steps that grows polynomially in the problem size. So if you happen to stumble across what you think is a solution to a problem in $NP$, you can verify or refute that it is indeed a solution in polynomial time. A good example is the problem of solving a jigsaw puzzle. If the puzzle has a large number of pieces, putting it together correctly is hard. But to check that any particular assembly is really a correct solution to the puzzle is easy: just look at it. Clearly, the polynomial-time problems are a subset of those in $NP$. Here are some examples of other problems that are also in $NP$:

- *Routing problem:* Suppose you are a salesperson who has to visit clients in a number of cities, and you want to make your round of visits without visiting any city more than once. Given the network of cities and the roads linking them, is there a route that starts and finishes at the same city and visits every other city exactly once?

- *Assignment problem:* Given information on lecture times, students, and courses, does there exist a timetable for each student that has no conflicts?

- *Map-coloring problem:* Does there exist a way to color a given map using just three colors, so that no two countries with a common border (greater than a single point) have the same color?

177

- *Bin-packing problem:* Given a collection of packages, together with their sizes, and a collection of identical boxes, does there exist an assignment of packages to boxes such that every package can be placed in a box without overflowing any of the boxes?

There are many problems in $NP$ for which it's unknown whether or not they are also in $P$. For example, the question of whether or not $n$ linear constraints on $n$ variables have a solution in integers is of great importance in optimization theory, as we'll see in the next chapter. Yet there is no known polynomial-time procedure for deciding this question. While we have as yet no ironclad proof showing that $P \neq NP$, most computer scientists would be shocked into a state of total catatonia if that turned out not to be the case. One of the main reasons is that a very large number of $NP$ problems have all been shown to be equivalent in the sense that if one of them turns out to be in $P$, then they are all in $P$.

This fact, which was proved by Stephen Cook in 1971, is probably the central result of computational complexity theory. It serves to motivate work in the field, since it says that in order to refute the $P = NP$ assertion, all we need do is produce a single instance of an $NP$ problem for which there cannot exist a polynomial-time algorithm for its solution. But so far no one has managed to find this elusive counterexample. Who knows, maybe the assertion $P = NP$ will turn out to be an undecidable proposition, thus independent of the usual axiomatic framework of mathematics, just as the famous Continuum Hypothesis was shown to be neither provable nor unprovable in the 1960s. Perhaps.

# Models of Computation

The universal Turing machine is a device for formalizing the idea of a computation over the integers: it starts with an integer input (a long string of 0s and 1s containing the describing the program $\mathcal{P}$ and the input data $\mathcal{I}$), producing a binary string (the output)—provided the program $\mathcal{P}$ eventually halts when processing the input $\mathcal{I}$. This is a *model* for what we mean when we speak of carrying out a "computation." And within the mathematical framework of this Turing-machine idealization of a real-world, pencil and paper computation, we have seen that some numbers are computable while many others are not. But other models of computation are possible.

In the mid-1980s, Berkeley mathematician Stephen Smale became interested in the foundations of numerical analysis and, as a result, began tormenting numerical analysts with the question, "What is an algorithm?" He reports that he usually did not receive a satisfactory answer to the question—with one exception. When asked to characterize an algorithm, Arieh Iserles replied, "A FORTRAN program." Taking this as his cue, Smale, along with his colleagues Lenore Blum and Mike Shub, developed a model of computation that can be seen as a mathematical idealization of a FORTRAN program.

The Blum–Shub–Smale (BSS) model of computation is a simple abstraction of the flowchart for a computer program. Figure 4.8, a flowchart for using Newton's Method with the real number input $x$ to find an approximation to a square root of a real number $c$, illustrates their basic idea. The essential ingredients of the BSS model for computation are all contained in this flowchart. They are:

1. A space of *inputs*. (Here, the real numbers.)
2. A space of *outputs*. (Here, again, the real numbers.)
3. A space of *states,* where the computation takes place. (In this case, the computation is simply the operation $x \to \frac{1}{2}(x + c/x)$.)
4. A distinguished set of inputs $\Gamma$, for which the BBS machine halts. We call this the *halting set* for the machine. (In this example, $\Gamma$ is simply the set of initial guesses $x$ that lead to an approximation to the square root of $c$ that is within a distance $\epsilon$ of the true value.)

It's easy to see that the flow of the program defines an input/output map $\phi$ from the set $\Gamma$ to the output set.

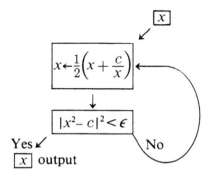

**Figure 4.8** Flowchart for calculating the square root of a number.

In this BBS setup, a *computable map* is a map like $\phi$ that is the input/output map of some machine $M$. It is an easy exercise to show that polynomial maps, rational maps, and many other elementary maps are computable, as are compositions of computable maps. It also turns out that the BBS setup over the reals can be extended to computation over a fairly general class of number systems, including the integers and the complex numbers. Moreover, Blum, Shub, and Smale have shown that the computable functions over the integers coincide with the traditional computable functions defined by the Turing machine. So the BSS machines constitute a legitimate generalization of the Turing notion of computation.

Computability deals with the black-or-white issue of whether a function (or number) can be calculated—even in principle. By introducing the polynomial/exponential dichotomy, we divided the class of computable functions into those that are computable by an efficient algorithm and those that aren't. Useful as this division is, it still deals with the matter of worst-case analysis of general functions. But in the everyday world of numerical analysis and computation we don't compute "general functions." Nor, by definition, do we typically face the worst case. So from a really *practical* point of view, what we're concerned with is the question of how hard it is to compute the solution to the average, or typical, case. This is part of the general theme that we take up in the next chapter, which considers problems of how to optimally allocate resources, computational or otherwise.

# CHAPTER

# 5

# The Simplex Method

*Optimization Theory*

# The Math of a Traveler

Suppose your Polly from Peoria wants to take a European tour during her summer vacation, and decides to visit Rome, Paris, London, Stockholm, and Vienna. If she knows the cost of an airline ticket between any two of these cities, what is the cheapest way for her to visit all the cities once and end up back where she started? Is it better for her to travel London-Paris-Rome-Vienna-Stockholm-London? Or would the tour Paris-London-Stockholm-Vienna-Rome-Paris be less expensive? Or is there yet another tour that's still cheaper? Finding the best possible route for Aunt Polly's vacation is the puzzle now known as the *Traveling Salesman Problem* (TSP).

At first glance, the solution to Aunt Polly's problem looks easy. She has five possible cities at which to begin her tour. After choosing one of them, she can then go to any one of the remaining four cities. There are then three cities to choose among for the next stop, and so on. Thus, Aunt Polly needs only to add up the city-to-city costs for each possible routing and pick the tour having the lowest cost. For this five-city case, there are only $5 \times 4 \times 3 \times 2 \times 1 = 120$ possible tours. A computer would probably take only a few nanoseconds to examine all these possibilities and tell Aunt Polly the best way to spend her travel dollar. Unfortunately, this straightforward approach involving an exhaustive examination of all possibilities is generally useless if the number of cities is even as large as, say, 100. In that case about $10^{150}$ calculations must be performed to look at all possible tours, requiring around $10^{120}$ years on today's fastest supercomputer!

While not many tourists or sales reps want or need to visit 100 cities, the Traveling Salesman Problem is important because it has far wider applicability than just in the travel industry. For example, electronic circuit board manufacturers have to drill as many as 65,000 holes on their boards using laser drills. Finding the best way to drill the holes turns out to be a TSP, since it involves finding the shortest tour that visits each hole position exactly once. And, in fact, a team of academic and industrial researchers recently established a record for the TSP by finding the exact minimal-distance path for visiting 3,038 cities, where the "cities" were indeed holes drilled on a printed circuit board. Just for fun, Figure 5.1 shows the solution to this problem.

183

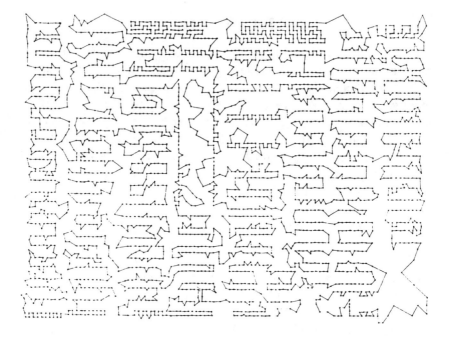

**Figure 5.1**    Solution of a TSP with 3,038 cities.

The manufacture of Japanese pachinko machines offers an even more complicated version of the TSP. These pinball-like machines have thousands of nails that the ball bounces off of, and it's a major problem deciding how to move the hammering head so as to insert these nails as quickly as possible. The problem is made even more difficult by the fact that the nails stand up higher than the hammer, so that the hammer cannot revisit any location at which it's already pounded-in a nail.

These kinds of TSPs are quintessential examples of what for our purposes are called "optimization problems." These are problems whose mathematical statement involves a set of variables, the values of which may be subject to constraints of one sort or another. Moreover, we assume a rule is given that assigns a number to each possible set of values of the variables. The goal is to find values of the variables that make this number as large (or as small) as possible, consistent with the requirement that these values of the variables satisfy the given constraints.

To illustrate the general idea, suppose we have a total of $1,000 to divide among two possible investments, say stocks and Treasury bills.

Further, assume that an investment of $X$ dollars in stocks produces an annual return of $2X$ dollars, while $Y$ dollars placed in T-bills yields $Y^2$ dollars (I never said this problem was realistic!). Since our interest lies in obtaining as large a return as possible, we want to select $X$ and $Y$ so as to maximize the quantity $2X + Y^2$, subject to the budgetary constraint $X + Y \leq 1,000$. So here the problem variables are $X$ and $Y$, the amount of money placed in stocks and T-bills, respectively, while the single constraint is simply that we can't put in more money than we have. Finally, any feasible allocation of money $X$ and $Y$ between the two investments is measured by the number $f(X, Y) = 2X + Y^2$. The job is to find an allocation $X$ and $Y$ that makes this the value of the function $f$ as large as possible, subject to the budgetary constraint. In this especially simple (and silly!) case, it's clear that the best place to put the money is to dump the entire amount into T-bills and allocate nothing to stocks, thereby obtaining a return of $(\$1,000)^2 = \$1$ million per year!

The investigation of optimization problems like the TSP began to enter the mathematical consciousness as a recognizable and respectable discipline during the latter part of the 1930s, gaining considerable visibility and attention during and immediately after the Second World War. At this time, many types of business and military questions involving things like the best way to schedule aircraft maintenance, allocate money to investments or process parts in an assembly-line operation came to be part of the field now called "operations research." Operations researchers are concerned with creating a minimal cost (or maximal benefit) schedule, or plan, of what is to happen and when. In the early days of operations research, such plans were labeled "programs." This led to the terminology that persists to this day, by which we call finding optimal schedules, like the best tour in a TSP, a "programming" problem. So despite the fact that the development of the digital computer was crucial for the practical implementation of the techniques needed to solve these types of optimization problems, a programming problem has nothing to do with computer programs, per se. Rather, it refers to the determination of a plan of action that is in some sense optimal.

Having settled this small bit of possible terminological confusion, let's look at the simplest, and what from a practical dollars-and-sense point of view is certainly the most important class of optimization problems, those for which everything in sight is linear.

## Thinking Linearly

Most management decisions ultimately come down to decisions about how to allocate resources so as to optimize something, for example, the allocation of money to maximize return on investment or the allocation of people and materials to minimize the total cost (and thereby maximize the profit) in producing a product like a car or a television set. Often these kinds of problems can be formulated in such a way that they can be solved by a procedure called *linear programming* (LP). But before describing this method in general terms, let's look at a very simple example just to get the general idea.

Consider the situation faced by the Chow-Down Dogfood Company, which manufactures two types of dog food, the Bow-Wow and the Wuff-Wuff brands. Both brands are mixtures of lamb, fish, and beef compounds, the exact mix being what distinguishes the two brands. Table 5.1 shows the amount of each of the compounds that's needed to produce one package of both Bow-Wow and Wuff-Wuff. Table 5.1 also shows the total amount of each compound that the company currently has in stock.

Assume the company makes a profit of $12 on each package of Bow-Wow and nets $8 per package of Wuff-Wuff. Then the problem for the Chow-Down Company is to decide how many packages of each brand to produce so as to maximize its total profit.

To formulate this problem mathematically, let $B$ be the number of packages of Bow-Wow that are produced, while $W$ denotes the production level of Wuff-Wuff. Then from Table 5.1 we see that the total amount of the lamb compound that will be used is $4B + 4W$ pounds. But since there are only 1,400 pounds of the lamb compound available, Chow-Down must obey the constraint

$$4B + 4W \leq 1,400, \tag{1}$$

which mathematically expresses the fact that Chow-Down can't use more lamb compound than they have at hand. Similar arguments based on the total amount of the fish and beef compounds in stock lead to the constraints

$$6B + 3W \leq 1,800, \tag{2}$$

$$2B + 6W \leq 1,800. \tag{3}$$

**Table 5.1**  Ingredients needed for a package of Chow-Down dog food brands.

| Compound | Total Amount Available | Amount in One Pkg. of Bow-Wow | Amount in One Pkg. of Wuff-Wuff |
|:---:|:---:|:---:|:---:|
| Lamb | 1,400 lb | 4 lb | 4 lb |
| Fish | 1,800 lb | 6 lb | 3 lb |
| Beef | 1,800 lb | 2 lb | 6 lb |

Since Chow-Down cannot produce a negative number of packages of either brand, we also require that $B$ and $W$ be nonnegative.

Note that these are all *linear* constraints, since the unknowns $B$ and $W$ appear to the first power everywhere. Finally, since the company earns a profit of $12 per package of Bow-Wow and $8 per package of Wuff-Wuff, the total profit to Chow-Down, call it $P$, can be mathematically expressed as

$$P = 12B + 8W. \qquad (\dagger)$$

The problem is then to find production levels of the two brands, that is, values of $B$ and $W$, that make $P$ as large as possible, subject to the constraints (1) through (3).

The easiest way to solve this problem is to draw a picture. Any pair of values of $B$ and $W$ constitute a point in the two-dimensional plane whose coordinates axes are $B$ and $W$. And since both quantities must be nonnegative, we can confine our attention to points in the first quadrant. Moreover, the three constraints are all linear. This means that each constraint can be geometrically pictured as a straight line in the $(B, W)$ plane. The overall situation is shown in Figure 5.2, where the constraints (1), (2), and (3) are shown as the lines labeled ①, ②, and ③ in the figure. The *feasible set,* consisting of all those points $(B, W)$ satisfying the constraints, is shown as the shaded region 0ABCD, while the *boundary* of the feasible set is formed by the points on the hatched portions of the constraint lines. Finally, let's note the *vertex* or *corner points,* which are the points where the line segments forming the boundary intersect.

The dashed lines in Figure 5.2 represent various constant levels of profit. So, for instance, the points on the dashed line labeled "$P = 1,200$" all represent production levels of the two brands that yield a profit of $1,200 for Chow-Down. Since the company wants to maximize its profit, it seeks to move the line $P$ parallel to itself outward from

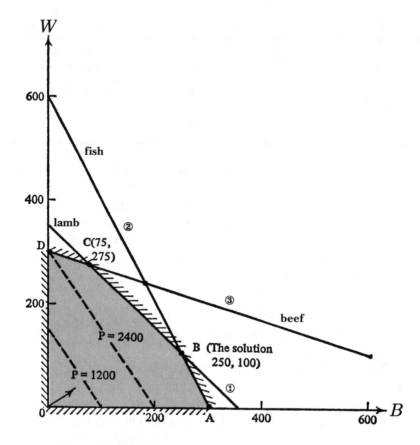

**Figure 5.2** Graphical solution of the dogfood mixing problem.

the origin as far as possible, subject to the condition that at least one point of the line intersects the boundary of the shaded region. A bit of experimentation shows that however the line $P$ happens to be situated, this parallel movement outward from the origin will end at one of the corner points on the boundary of the shaded region. Thus, the maximal profit must lie at one of the corner points—in this case one of the points 0, A, B, C, or D. As indicated in the figure, this turns out to be the point B, which has the coordinates $B = 250$, $W = 100$. The profit at this point is $3,800. So if Chow-Down wants to get as much benefit as possible out of their supply of the lamb, fish, and beef compounds, they should produce 250 packages of the Bow-Wow brand and 100 packages of Wuff-Wuff.

This example illustrates the most important feature characterizing any LP problem: the solution is always found at a corner point on the boundary of the feasible region. But the boundary points are where at least two of the problem constraints are up against their limits, that is, are strict equalities. Therefore, at a solution point at least two of the resources—lamb, beef, or fish—will be totally used up. We'll come back to this observation in a moment.

So we see that despite the fact that every point of the feasible set is, in principle, a candidate for the optimal solution, we really need only examine the corner points to find the optimum. The computational implications of this result are enormous, since the set over which we have to search for the solution is reduced from an infinite set (the set of all points in the shaded region) to a finite set (the set of all corner points). Of course, this may still be a difficult computational problem. After all, the number of possible chess positions is also finite, but far beyond the capacity of any real or imagined computer to ever examine in a time less than several lifetimes of the universe. Fortunately, though, the LP case is not quite this bad, and there exist computational procedures for searching the corner points in a computationally efficient and practical manner. We will now look at the first of these procedures, the so-called *Simplex Method,* which still forms the basis of most algorithms used in practice to solve linear programming problems.

## The Simplex Method

In 1947, George B. Dantzig was working as a civilian in the Pentagon as a mathematical adviser to the Air Force Comptroller. As part of his job, Dantzig was often called upon by the Air Force to solve real planning problems, involving ways to distribute Air Force personnel, money, planes, and other resources in a cost-effective fashion. Since most of these problems involved economics in one way or another, Dantzig enlisted the advice of economist Tjalling Koopmans about how to solve these LP problems, assuming that economists had developed solution techniques for them years before. To Dantzig's great surprise, Koopmans told him that economists didn't have any procedures for systematically finding the solution to LP problems either. So in the summer of 1947 Dantzig set out to find one.

The first—and most important—step in Dantzig's search for a method to solve LP problems was the observation we made about the dogfood mixing problem, namely, that the feasible region is what's called a *polytope*—a set like that shown in Figure 5.3. So by the same line of argument used in the two-variable problem of dogfood mixing, this means that the optimal point has to be at one of the corners of this set. Moreover, Dantzig argued, the criterion function generally has different values at each corner point. Therefore it should be possible to start at a arbitrarily chosen corner point at which the criterion function has a certain value, and improve the values of the criterion function by moving from such a corner point to an adjacent one, much like the movements made by beetle shown in the figure as it crawls along the edges seeking, say, the point containing the greatest amount of food, marked here by the chocolate cake. In the jargon of algebraic topology, a polytope of this sort is termed a "simplex," which gives rise to the name of Dantzig's algorithm telling the beetle how it should move along the edges to reach its goal.

Originally, Dantzig thought such a procedure might be hopelessly inefficient, wandering along improving edges from corner point to corner point for a long time before reaching the corner point at which the criterion function has its greatest value. But he was wrong! In fact, he discovered that in almost all cases finding the solution to the problem involved only as many moves as there are constraints in the problem. It should be pointed out, however, that this is an *empirical* observation

**Figure 5.3**  A beetle crawling along the edges of a polytope.

based on the actual solution of thousands of real-life problems, and it is possible to concoct special types of LP problems that do actually take a very long time to solve using the Simplex Method or that lead to degeneracies in the method so that it never finds the solution. But these kinds of problems are more like mathematical curiosities, and seldom, if ever, occur in practice. This happy fact leads to the fundamental result of optimization theory:

---

**THE SIMPLEX METHOD**  *Almost every linear programming problem can be solved by the following procedure:*

1. *Find a vertex representing a feasible solution (that is, one whose coordinates satisfy all the constraints of the problem), and calculate the value of the criterion function at that point.*
2. *Examine each boundary edge of the feasible set passing through this vertex to see whether movement along such an edge will improve the value of the criterion function.*
3. *If it does, move along an edge to the vertex point that yields the largest improvement in the current value of the criterion function.*
4. *Repeat steps (2) and (3) until there is no longer an edge along which movement improves the criterion function. The current vertex is then a solution to the problem.* ■

So we see that the computational implementation of the Simplex Method involves two aspects: (1) finding a feasible solution to start the process, and then (2) improving a feasible solution by moving along edges of the boundary of the feasible set from vertex-to-adjacent vertex, one step at a time, until reaching the optimal point.

To see how the Simplex Method actually works in practice, let's go back and use it to solve the dogfood mixing problem. Recall that problem involves finding nonnegative values of $B$ and $W$ that maximize the criterion function (†) given above, subject to the constraints (1) through (3) on the amount of lamb, beef, and fish compound available. Since inequalities are awkward to deal with, we introduce the additional nonnegative variables $s_1$, $s_2$, and $s_3$, called *slack variables,* whose purpose is to convert the inequality constraints into equalities. For example, if the values of $B$ and $W$ are such that $4B+4W$ is less than 1,400, the slack variable $s_1$ for constraint (1) takes on the value $s_1 = 1,400 - 4B - 4W$, which then makes $4B + 4W + s_1 = 1,400$. In what follows, the slack variables are always used in this fashion to ensure that constraint inequalities are converted to equalities.

Our problem can then be expressed as: Find values of the variables $B$ and $W$ that maximize

$$12B + 8W,$$

subject to the constraints

$$4B + 4W + s_1 = 1,400,$$
$$6B + 3W + s_2 = 1,800,$$
$$2B + 6W + s_3 = 1,800.$$

As always, we also have the constraints that the variables $B$, $W$, $s_1$, $s_2$, and $s_3$ be nonnegative.

At this point it's helpful to introduce a bit of terminology. A set of values of $B$, $W$ and the slack variables satisfying the constraints is called a *feasible solution*. A solution obtained by setting some of these variables equal to zero, thus leaving a set of, say, $m$ equations in $m$ variables, is called a *basic solution* and the set of $m$ nonzero variables is called a *basis*. It is a mathematical fact that if a system has a feasible solution, then it also has a basic feasible solution. Normally, this is the kind of solution we will be looking for. Now let's see how the Simplex Method works for the dog food problem.

The problem the Chow-Down Company has to solve has five variables—$B$, $W$, $s_1$, $s_2$, and $s_3$—together with the three constraint equations

$$4B + 4W + s_1 = 1,400,$$
$$6B + 3W + s_2 = 1,800,$$
$$2B + 6W + s_3 = 1,800.$$

So to form an initial basic solution we must choose three of the variables and set the others to zero. If the three variables we choose satisfy the constraint equations above, then we will have found a first basic feasible solution with these three variables as the basis. In order to make this initial choice, we note that each of the slack variables appears in only one of the constraint equations. This suggests using the slack variables to form the initial basic solution. This means setting the variables $B$ and $W$ equal to zero, which leaves us with a set of $m = 3$ constraint equations in the three unknowns $s_1$, $s_2$, and $s_3$. The solution to this set of equations can be easily read off by inspection as $s_1 = 1,400$, $s_2 = 1,800$ and $s_3 = 1,800$. The next step is to solve the constraint equations for the basic variables $s_1$, $s_2$, and $s_3$ in terms of the nonbasic ones, $B$ and $W$, and to write the criterion function in terms of the nonbasic variables.

Carrying out these operations, the values of $s_1$, $s_2$, $s_3$, and $P$ are

$$s_1 = 1,400 - 4B - 4W,$$
$$s_2 = 1,800 - 6B - 3W,$$
$$s_3 = 1,800 - 2B - 6W,$$
$$P = 12B + 8W.$$

Geometrically, this solution corresponds to the point 0 in Figure 5.2 shown in the section "Thinking Linearly" above; unfortunately, it also corresponds to a point at which the profit is 0, since both of the quantities $B$ and $W$ are out of the basis and thus have the value 0.

To increase the profit $P$, we can increase either $B$ or $W$. Since only one variable is changed at each step of the Simplex Method, it is the custom to choose that variable having the largest coefficient in the criterion function to introduce into the new basis. The reason is that this choice will increase the profit at the fastest possible rate. So we choose to introduce the variable $B$ into the basis. But we can't let $B$ be larger than 300, since that would make the slack variable $s_2$ negative (recall that $W = 0$ at this point). So we set $B = 300$, which makes $s_2 = 0$. We are now at the point $A$ in the figure, $B$ has become a basic variable and $s_2$ is a nonbasic one.

We again express the basic variables (which are now $s_1$, $B$, and $s_3$) and the criterion function for the profit $P$ in terms of the nonbasic variables $s_2$ and $W$, obtaining

$$s_1 = 200 + \tfrac{2}{3}s_2 - 2W,$$
$$B = 300 - \tfrac{1}{6}s_2 - \tfrac{1}{2}W,$$
$$s_3 = 1,200 + \tfrac{1}{3}s_2 - 5W,$$
$$P = 3,600 - 2s_2 + 2W.$$

Since the variables $s_2$ and $W$ on the right-hand side of this set of equations are zero at this stage, the constants in the above equations give the values of the basic variables, as well as the value of the criterion function. This means that by introducing $B$ as a basic variable, the profit has increased from 0 to 3,600. But we can still do better, since the nonbasic variable $W$ has a positive coefficient. This means that the profit $P$ can be further increased by introducing $W$ into a new set of basic variables.

As before, we note that the new basic variable $W$ cannot be greater than 100 since a larger value would make the slack variable $s_1$ negative. So we set $W = 100$ packages of Wuff-Wuff, and replace the slack variable $s_2$ by $W$ in our new basis. This leaves us with $B$, $W$, and $s_3$ as the new basis variables. Proceeding as above, we find the new values of the basic variables to be

$$W = 100 - \tfrac{1}{2}s_1 + \tfrac{1}{3}s_2,$$
$$B = 250 + \tfrac{1}{4}s_1 - \tfrac{1}{3}s_2,$$
$$s_3 = 700 + \tfrac{5}{2}s_1 - \tfrac{4}{3}s_2,$$
$$P = 3{,}800 - s_1 - \tfrac{4}{3}s_2.$$

Since both $s_1$ and $s_2$ have negative coefficients in the expression for $P$, an increase in either would only make the profit smaller. Thus we have reached the optimal solution, which corresponds to the point B in Figure 5.2. And since the variables on the right-hand side of the above set of equations all have the value 0 at this stage, we find the optimal solution by reading off the constants in these equations. Therefore, the optimal solution is for Chow-Down to produce 250 packages of Bow-Wow and 100 packages of Wuff-Wuff, which will produce a profit of $3,800. The fact that the slack variable $s_3 = 700$ in the optimal solution means that the third inequality constraint, the one for beef, is *inactive*. This implies that 700 pounds of the beef compound is not used in the optimal mix of brands. On the other hand, both the slack variables $s_1$ and $s_2$ are zero at the optimal point, telling us that Chow-Down must use all of its stock of the lamb and fish compounds in making up the optimal production levels of their two brands.

# Duals and Diets

One of the first major books outlining the importance of linear programming in economics was published in 1958 by the future Nobel Laureates Paul Samuelson and Robert Solow, along with their colleague Robert Dorfman. In this volume, the LP method was illustrated by what has come to be called the Diet Problem. It goes as follows.

Suppose a person requires a certain amount of two vitamins, $V$ and $W$, each day. These vitamins are found in two different foods, say milk and eggs. The daily requirement for the two vitamins, the amount of each vitamin in a unit amount of these foods, and the unit cost of the two foods (in cents) is given in the following table:

| Vitamin | Amount in Milk | Amount in Eggs | Daily Requirement |
|---------|----------------|----------------|-------------------|
| $V$ | 2 | 4 | 40 |
| $W$ | 3 | 2 | 50 |
| **Unit Cost** | 3 | $2\frac{1}{2}$ | |

The objective is to determine how much milk and eggs should be eaten per day to obtain the minimal daily requirement of vitamins at the lowest possible cost of purchasing these foods.

Letting $a$ represent the amount of milk and $b$ the amount of eggs to be purchased, the Diet Problem can be mathematically written as

$$\text{minimize } 3a + \tfrac{5}{2}b,$$

subject to the conditions

$$2a + 4b \geq 40,$$
$$3a + 2b \geq 50,$$
$$a, b \geq 0.$$

I'll leave it to the reader to use the techniques just presented to show that the solution to this problem is $a^* = 15$, $b^* = 5/2$, with the minimal cost being $51\frac{1}{4}$ cents. This formulation of the problem emphasizes the perspective of the buyer, who wishes to minimize overall food costs in acquiring the necessary vitamins. But there is another, equally valid, way of looking at this problem—from the perspective of the seller.

Consider the grocer who sells the milk and eggs to the person who requires the vitamins. The grocer knows that the foods have a certain value on account of their vitamin contents $V$ and $W$. The grocer's problem is to determine selling prices, say $X$ cents per unit of vitamin $V$ and $Y$ cents per unit of vitamin $W$. But the grocer is constrained in setting these prices by the fact that he can't set them higher than the prevailing market prices for milk and eggs. In other words, the price the grocer sets for milk cannot be greater than 3 cents per unit, nor can the price for eggs exceed $2\frac{1}{2}$ cents per unit, since otherwise the grocer will lose his customer in a competitive market. At the same time, however, the grocer wants to maximize the store's total income, which will be $40X + 50Y$, since the daily requirement is 40 units of vitamin $V$ and 50 units of vitamin $W$. The grocer's problem can be stated mathematically as

$$\text{maximize } 40X + 50Y,$$

subject to the conditions

$$2X + 3Y \leq 3,$$
$$4X + 2Y \leq 5/2,$$
$$X, Y \geq 0.$$

Comparison of the buyer's problem and the grocer's problem turns up the remarkable fact that the first can be transformed into the second by making the following substitutions:

$$\text{minimize } \rightarrow \text{ maximize}$$
$$\geq \quad \rightarrow \quad \leq$$
$$\text{food costs } \rightarrow \text{ price constraints}$$

The problems of the buyer and grocer are called *dual LP problems,* and the above substitutions show that every LP problem contains two completely equivalent formulations: the so-called *primal* problem, which is the problem as originally stated, and its *dual* problem, which is formed by using the set of substitutions just given.

Mathematically, this duality between the primal and the dual problem is of exactly the same character as the duality between points and lines in ordinary euclidean geometry. In that setting, we know that every statement that Euclid makes about the relationship between points can be replaced by a completely equivalent statement about lines. For instance, the statement "two points determine a line" has the dual statement "two lines determine a point," which is obtained simply by interchanging the words "point" and "line." Mathematics is filled with such dualities, with the one given here for the buyer and grocer being the manifestation of this Duality Principle in the specific context of linear programming.

The examples worked out here show all the aspects of the Simplex Method as it is generally applied in practice. But the reader desiring more details on how to implement the procedure computationally, as well as discussion of more technical aspects like duality, should consult the many works cited in the chapter Bibliography. Now let's look briefly at a point that was glossed over in the foregoing discussion, namely, the implicit assumption that the variables in an LP problem can assume *any* nonnegative values whatsoever.

## Integer Programming

The proverbial perceptive reader will have noticed that in the dogfood-mixing problem we took no explicit account of the fact that packages of dog food come in integral units. In other words, Chow-Down can't produce one-third or 12/79ths of a package of one of their brands; the company can produce only 0, 1, 2, or some other integer number of packages. And it was only by a fortuitous choice of the various coefficients in the constraints that the optimal solution to the problem happened to turn out to be in integers. With a bit of experimentation, the adventurous reader will soon discover that it doesn't take much of a change in the constraints to have all the corner points lie at nonintegral values of $B$ and $W$, so that the Simplex Method will generate an optimal solution that is nonintegral. There are a large number of problems of practical concern that have the feature that the problem variables only make sense when they take on integer values. Examples of this include the allocation of money to various investments, the assignment of students to lecture rooms and the scheduling of assembly operations in a factory.

At first glance, one might think the way to deal with the integer-valued constraint is to simply to solve the problem without this constraint, and then round off the solution that's obtained to its nearest integer levels. Geometrically, the reason this simpleminded approach doesn't work is that the corner point representing the solution to an LP problem is not a continuous function of the coefficients that appear in the criterion function and constraint equations defining the problem. Therefore, a small change in any of these coefficients can cause a "jump" in the solution from one corner point to another. And rounding off a non-integral solution is mathematically tantamount to "jiggling" some of the coefficients. Here's a small example showing the folly of the rounding-off approach to these types of integer programming problems.

Consider the problem of maximizing

$$11x - y$$

subject to the constraints

$$10x - y \le 40,$$
$$x + y \le 41/2.$$

Here the maximization is taken over all nonnegative *integer* values of the unknowns $x$ and $y$.

If we ignore the integer requirement and apply the Simplex Method, we obtain the optimal solution $x = 11/2$, $y = 15$, which leads to a value of the criterion function of $91/2$. But since $x$ is not an integer, we cannot accept this solution. The nearest integers to $x$ are 5 and 6, so we might think of testing the integer pairs $x = 5$, $y = 15$ and $x = 6$, $y = 15$ as possible solutions to the problem. The first pair gives a criterion value of 40, while the latter pair is not even feasible since it doesn't satisfy the problem constraints. As it turns out, there is a better integer pair for this problem, namely, $x = 5$, $y = 10$, which leads to a value of 45 for the criterion function.

This example shows that integer programming problems need to be treated on their own merits, and cannot be dismissed by the rule: "Round off the solution from the Simplex Method to the nearest integer." For an account of special methods developed by R. E. Gomory and others to treat this vitally important class of optimization problems, we refer the reader to the material cited in the bibliography.

While we could devote another book several times the size of this one to various extensions and generalizations of LP, space considerations allow only a small dip into this ocean. So to conclude our discussion of linear optimization problems, we consider perhaps the most important special class of questions that can be dealt with using LP-based ideas, flows in networks.

## Graphs and Bridges

Two hundred and fifty years ago, the modern Russian city of Kaliningrad was part of what was then the German territory of East Prussia, and the city was called Königsberg. A popular Sunday afternoon pastime for the residents of Königsberg was to stroll through the center of town, which is bisected by the River Pregel. At that time there were seven bridges crossing the river, whose positions are shown in the left side of Figure 5.4. It eventually became a bit of a local puzzle to ask if there existed a path crossing each bridge exactly once. The great mathematician Leonhard Euler heard about this problem in 1736 and saw immediately how to solve it.

Euler's key insight into solving the problem of the Königsberg bridges was to realize that the size and shape of the land masses bounded by the water, as well as the length of the bridges, play no role whatsoever in the question. So he reduced the problem to the graph shown on the right-hand side of Figure 5.4, in which the nodes of the graph represent the land masses, while the edges connecting the nodes stand for the bridges. In other words, Euler reduced the land masses to points and the bridges to lines. He then reasoned that a node could be one of three types: (a) the beginning of a path, (b) the end of a path, or (c) an intermediate node. If the latter, then the path must both enter and exit the node; hence, such a node must have as many edges coming into it as going out. Therefore, the total number of edges incident on the node must be an even number, that is, the node has even degree. On the other hand, if a node is

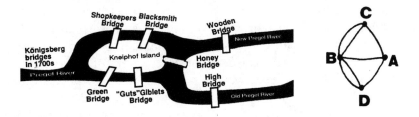

**Figure 5.4**   The seven bridges of Königsberg.

either the beginning or the end of a path—but not both—there must be an odd number of edges incident upon it, namely, one. And since any path traversing each bridge exactly once must have a beginning and an end, Euler concluded that any such path would have exactly two nodes of odd degree. But inspection of the right side of Figure 5.4 shows that in the Königsberg graph all four nodes have odd degree. Thus, there cannot exist a path that goes across each bridge exactly once. Incidentally, only three of the original seven bridges still remain (Honey Bridge, Wooden Bridge, and High Bridge) and there is now a bridgeway that passes completely over the island. So the original problem is no longer a problem—and so much the worse for mathematics.

Euler's treatment of the problem of the Königsberg bridges marked the beginning of what we now call graph theory. An abstract *graph* $G$ is simply a set $V$ of *nodes,* or *vertices,* together with a set $E$ of *edges,* or *arcs,* connecting various pairs of elements from $V$. Thus, if there is an edge going from node $i$ to node $j$, then the pair $(i, j)$ is in $E$. If each edge has a direction (that is, a beginning and an end), then we call $G$ a *directed graph,* or *digraph.* Thus, for a digraph we may have $(i, j)$ in $E$, while $(j, i)$ is not in $E$.

A graph is a good way to mathematically represent a physical situation in which there is a flow of something—materials, people, money, information—from one place to another. For example, a food distribution system could be represented by a graph whose nodes are the various cities to be supplied from others, with some nodes being the sources of the food (warehouses, for example) and the others being destination cities. The edges of such a graph would then represent the road/rail/air links that exist between the various pairs of cities. Such a combination of source nodes, sink nodes, intermediate nodes and links is often termed a *network.*

One of the most important questions we can ask about a network that is mathematically represented by such an abstract graph is how some commodity like information, material, money, and/or people can flow optimally from one part of the network to another. Suppose we single out a particular node in the graph, calling it the *source* of the commodity that is flowing through the network to some other particular node, which we call the *sink.* The question that we'll concern ourselves with now is: What is the fastest steady-state flow at which the commodity can move from the source to the sink?

## And So It Flows

The crucial concept we need to solve the network flow problem is the idea of a *cut*. This is simply a separation of the nodes of the graph into two sets $P$ and $\bar{P}$ having no common element, and such that the source node(s) is in one of the sets while the sink node(s) is in the other. To illustrate the idea, the simple five-node graph of Figure 5.5 shows a cut in which $P = \{s, p, q\}$, while $\bar{P} = \{r, t\}$.

In order to determine the maximum steady-state flow that can move through the network, we need to talk about the capacity of a cut. Suppose we have a cut $(P, \bar{P})$. We define the capacity of this cut, $c(P, \bar{P})$, to be the sum of all the capacities along all edges that have one node in $P$ and their other node in $\bar{P}$. To illustrate, the capacity of the cut shown above in Figure 5.5 is given by the sum

$$c(P, \bar{P}) = c(s, r) + c(q, r) + c(p, t) = 3 + 4 + 8 = 15.$$

By definition, a *minimal cut* is a cut of minimal capacity. Again using the graph of Figure 5.5, the minimal cut consists of the pair $P = \{s, q, r\}$, $\bar{P} = \{p, t\}$, which has capacity 13. Now we are in a position to state the principal result for network flows.

Let $V^*$ be the value of the maximal steady-state flow between the nodes $s$ and $t$, and let $c(P^*, \bar{P}^*)$ be the capacity of the minimum cut separating $s$ and $t$. Then we have the celebrated

---

**MIN CUT-MAX FLOW THEOREM**   $V^* = c(P^*, \bar{P}^*)$.   ∎

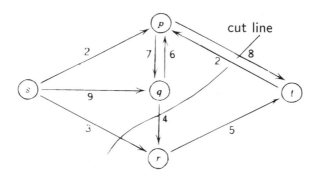

**Figure 5.5**   A cut in a graph.

This theorem was proved by Lester Ford and D. R. Fulkerson in the mid 1950s. It says that finding the maximal flow through the network is equivalent to finding a cut of minimal capacity. The reader might suspect, and rightly so, that the quantities $V^*$, $P^*$ and $\bar{P}^*$ could be found by standard linear programming methods. That this is indeed the case is borne out by a quick glance at the material cited in the bibliography. For now, let's fix these ideas of cuts and flows in place with a simple highway traffic example.

### Highway Traffic Flow

Consider the network shown in Figure 5.6, which we can regard as the roads in a freeway network. Suppose there is some roadwork going on, so that a detour begins at node $s$ and ends at node $t$. The flow capacity measured in, say, hundreds of cars per hour, of each segment of road in the detour network is given by the numbers not circled in the figure. Our task is to determine how to set up the detour route so as to maximize the traffic flow through the network. The Min-Cut, Max-Flow Theorem shows us how to do this.

We must examine all the possible cuts $(P, \bar{P})$ separating the beginning of the detour (node $s$) and its end (node $t$). Some of these cuts and their capacities are shown in Table 5.2. From the table, we see that the minimal cut has a capacity of 8. This means that $V^* = 8$, with $P^* = \{s, a, b, c, d\}$ and $\bar{P}^* = \{t\}$. Using this information, we can start at the sink node $t$ and trace back the optimal flow along each edge of the network, which is indicated by the numbers in black circles in Figure 5.6. Of course, for a larger network this "hands-on" approach to calculating the solution is rather unwieldy, and we need to make use of

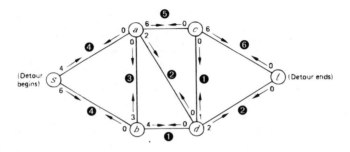

**Figure 5.6**   Highway detour network.

**Table 5.2** Cuts and capacities.

| $P$ | $\bar{P}$ | $c(P, \bar{P})$ |
|:---:|:---:|:---:|
| $\{s, a, b, d\}$ | $\{c, t\}$ | 9 |
| $\{s, a, b\}$ | $\{c, d, t\}$ | 12 |
| $\{s\}$ | $\{a, b, c, d, t\}$ | 10 |
| $\{s, a, b, c, d\}$ | $\{t\}$ | **8** |
| $\{s, a\}$ | $\{b, c, d, t\}$ | 14 |

more systematic procedures based on linear programming algorithms. The reader can find more details about how to do this in the material cited for this chapter in the bibliography.

To this point our attention has centered on linear problems and linear methods, principally linear programming and its extensions. But important as such problems are, the fact is that we live in a nonlinear world. Economies of scale in, say, production of dog food means that costs do not increase in direct proportion to the number of packages produced. Similarly, the amount of electricity you can push through a transmission network does not double if you double the capacity of the lines, but rather goes up more slowly due to various resistance and heating factors. These are nonlinear effects, reflecting various laws of physics and economics. So the time has come to shift gears and tackle the way the world really is, not how we might like it to be.

## The Welfare of the Masses

Earlier in this century, the architects of the Modern Movement based their work on a combination of science and art to create buildings that were designed to be the simplest, cheapest, and most direct answer to the problem of housing shortages. One of the preoccupations of the Modern Movement was the welfare of the masses, and in a 1931 address Walter Gropius, one of the leaders of the Movement, spoke to the question of how to best lay out housing blocks on a site. Gropius argued that the essentials for a wholesome life are light, air, and space, above and beyond adequate food and warmth. He went on to dismiss the then-popular city-block layouts for mass housing, advocating instead simple parallel blocks of housing. These contrasting styles are displayed in Figure 5.7.

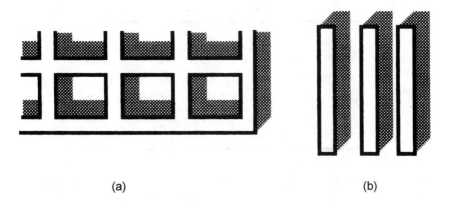

(a)                                            (b)

**Figure 5.7**    (a) City-block layout. (b) Parallel-block layout.

As part of Gropius's call to arms, he claimed that high-rise buildings of 10 to 12 stories are better than shorter 3- to 5-story buildings, because a higher population density is then possible at less cost while still preserving "light, air, and elbow room." In particular, Gropius claimed that the amount of open space for each inhabitant increases as the height of the block increases. To support this argument, he put forth three rules governing parallel housing blocks. Each rule states how the number of stories in a block varies with either the population in a block, the site area occupied by the block, or the angle of incidence of sunlight to the bottom floor of each block. Use of Gropius's rules to test his claim that higher blocks are better than lower ones insofar as giving each residence more "elbow room" leads to a nonlinear programming problem. Here are Gropius's three rules:

- **a.** Given a certain site area and sunlight incidence, the number of people housed increases with the number of stories.
- **b.** Given a certain sunlight incidence and a number of people to house, the size of the required site decreases with increasing number of stories.
- **c.** Given a site area and a number of people to house, the sunlight incidence decreases with increasing number of stories.

Gropius believed that these rules implied that the optimal height of buildings should be between 10 and 12 stories. In the late 1930s, several urban housing projects were built on this basis, and in the reconstruction after World War II many governments throughout Europe designed mass housing that incorporated Gropius's rules. But was he right? Let's turn to optimization theory and see.

There are three independent variables in Gropius's rules:

$P$ : The number of people to be housed.

$A$ : The site area.

$I$ : The tangent of the angle of incidence of the sun.

These variables affect the dependent variable $x$, the number of stories in each block of housing. In addition, we have the parameters and the constant

$a$ : The width of each block.

$b$ : The floor area per inhabitant.

$\ell$ : The length of each block.

$s$ : The distance between blocks.

$3$ : The height of each story (which is 3 meters).

All these variables and parameters are shown in Figure 5.8.

To see whether or not Gropius was correct in his belief that buildings of 10 to 12 stories give optimal space, we work with the site-area ratio (SAR), which is simply the area per person, or $A/P$. This is clearly inversely related to the population density. Using the fact that

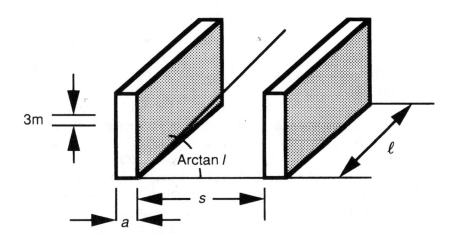

**Figure 5.8** Variables used in Gropius's housing model.

$$P = a\ell x/b,$$
$$A = \ell(a + s),$$
$$I = 3x/s,$$

we find that

$$\frac{A}{P} = \frac{b(aI + 3x)}{a\ell x}.$$

Gropius's claim is that for a constant population density (or SAR), the open space per person increases with the number of stories, and reaches its maximum at approximately 10 to 12 stories.

The amount of open space is given by the length of each block times the distance between blocks, or $s \times \ell$. Therefore, to obtain the amount of open space per person, or the "open-space ratio (OSR)," all we have to do is divide this quantity by $P$, the number of people to be housed, yielding $OSR = s\ell/P$. But we can substitute the expression for $P$ (given a moment ago as $P = a\ell x/b$) into this formula to obtain

$$OSR = \frac{bs\ell}{a\ell x} = \frac{bs}{ax}.$$

Since the SAR is defined as $A/P$, this leads to the relation

$$SAR - \frac{b}{x} = \frac{A}{P} - \frac{b}{x} = \frac{1}{P}\left(A - \frac{bP}{x}\right)$$

$$= \frac{1}{P}\left(a\ell + s\ell - \frac{ba\ell x}{bx}\right) = \frac{1}{P}s\ell = OSR.$$

Holding the population density (that is, SAR) constant, we see that indeed the OSR will increase *nonlinearly* with an increase in the number of stories $x$. Furthermore, the maximum OSR equals the population density and occurs when the number of stories $x$ becomes infinite. So Gropius is correct in his first assertion that the higher the block, the greater the OSR. But this isn't the whole story. We still have to examine his claim that 10 to 12 stories is the optimal height for buildings.

To do this, let's put some realistic numbers into these formulas. For values of the SAR, we consider current population densities. These appear to range from about 54 ft$^2$ per person in the very worst tenements of Hong Kong to around 5,400 ft$^2$ per person in sprawling cities like

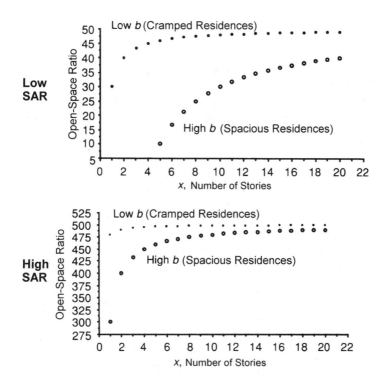

**Figure 5.9** The OSR of housing blocks at high and low values of SAR and *b*.

Sydney or Los Angeles. Figure 5.9 shows the OSR achieved for both low and high SARs. What this figure tells us is that at very high population densities (low SARs) and a low value of *b* (that is, very cramped quarters), the OSR is very near its maximum by about six stories. On the other hand, for very high values of floor area per inhabitant *b* (that is, spacious residences), then even at 20 stories the OSR is nowhere near its maximum. But if the goal is a low-density city, with a high OSR, the figure shows that the maximum is approached with housing units of around six to eight stories for both cramped and spacious residences. Thus, Gropius's conclusions seem fairly sound when applied to high-density cities with spacious residences. But in all other cases it seems that it would be much better to aim for housing blocks that are about half as high as what Gropius recommended.

Gropius's housing problem is an example of an *unconstrained* non-linear optimization problem, since the relation between the open-space ratio $OSR$ and the number of stories $x$ is not linear but rather is one of inverse proportion. Most optimization problems involve an allocation of some sort of resource—energy, money, materials—that is in limited supply; hence, there are resource constraints that must be satisfied by any solution. So our task for the next few pages is to outline a few of the ideas upon which methods are based for solving problems in which there is a nonlinear criterion function and/or nonlinear constraints.

## Hill Climbing

Think about the plight of the hiker Stuart, who's climbing a local hill. Stu has a map and compass. But as he makes his way up the hill a mist descends, blocking his view of everything but the immediate vicinity. So how should he proceed? On the assumption that he would like to get to the top of the hill as quickly as possible, one good policy would be for Stu to start walking straight up the hill. That is, go in the direction of steepest ascent. Consulting his map, Stu knows that this direction is at right angles to the contour line passing through his current location. The rate at which Stu's altitude increases in this direction of steepest ascent is called the *gradient*. And it's clear that if there are no subsidiary peaks in the local area, he will eventually get to the top by always moving in the direction of the gradient. The general path followed by Stu in this kind of situation is like the path A-B-C-D-S depicted in Figure 5.10.

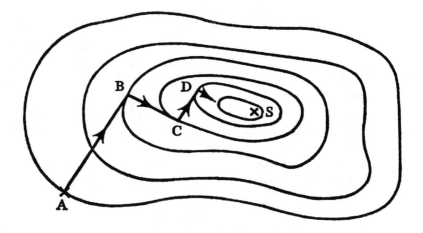

**Figure 5.10** Climbing a hill by following the gradient.

For solving nonlinear optimization problems, we can mimic the procedure used by Stu in getting to the top of the hill. Think of the value of the criterion function $f$ as being Stu's "altitude," with the values of the problem variables representing his "position." We calculate the gradient of the criterion function at his current position, the point $x^*$, which tells Stu the direction of maximum increase in $f$ from that point. So if there are no constraints, the general algorithm for the *method of steepest ascent*, or the *gradient method*, is as follows:

1. Start at an arbitrary point $x_0$;
2. Calculate the gradient of $f$ at $x_0$;
3. Move to the new point $x_1$ by taking a step of size $\delta$ in the direction of the gradient, where $\delta$ is a small positive constant;
4. Repeat steps (2) and (3) until a point $x^*$ is reached at which the gradient of $f$ is zero.

The foregoing procedure produces a trajectory much like Stu's climb in Figure 5.10. A serious difficulty with the method, however, is that it is purely local in the sense that it always ends up at the peak nearest to the starting point. So unless the criterion function $f$ has some special property ensuring that it has only a single local maximum, we can never be sure that the point at which the gradient method stops will be the true *global* maximum. Therefore, by following the gradient a mountain climber may get stuck at the top of a nearby hill or ridge instead of reaching the peak.

An obvious way to deal with this difficulty is to employ the method from several randomly selected starting points, and then pick the one that yields the highest value from among the points where the method of steepest ascent stops. But using this procedure, too, we can run into trouble, since in a problem with several hundred, or even thousands, of variables it may require a very large number of starting points to get a representative sample of possible paths. This, in turn, may lead to unacceptable amounts of computing time in following the gradient procedure to its termination from all these different starting points. Moreover, the structure of the criterion function may work against us, in the sense that it may have "ridges" along which the method of steepest ascent gets stuck. In such cases, we need special techniques to get the process "unstuck" and back on track again. The reader can find an account of these matters in the bibliography.

It remains an unhappy fact that there is no best method for finding the solution to general nonlinear optimization problems. About the best general procedure yet devised is one that relies upon imbedding the original problem within a family of problems, and then developing relations linking one member of the family to another. If this can be done adroitly so that one family member is easily solvable, then these relations can be used to step forward from the solution of the easy problem to that of the original problem. This is the key idea underlying *dynamic programming,* the most flexible and powerful of all optimization methods. Dynamic programming had its origin in the work of statisticians like Abraham Wald, who studied decision procedures for determining how many items had to be sampled in order to meet a given quality control level. But it was not until the early 1950s that Richard Bellman showed the great generality of the underlying idea, employing dynamic programming to address a dazzling array of questions arising in optimal control theory, game theory, production and scheduling processes, and much more. Let's see how it works by examining some problems.

# Routing in Networks

Flow problems involve the passage of commodities through a network. But often our concerns center on the passage of people or information rather than commodities. Probably the most common situation of this type is when we want to move from one part of a network to another at minimal cost—either in time, money, energy, or some other resource. This leads to a consideration of what are called *routing problems.* Here's a simple example.

### A Four-City Airline Problem

Consider an airline serving four cities, where the travel time (in hours) from city $i$ to city $j$ is the entry in row $i$, column $j$ of the matrix

$$T = \begin{array}{c c} & \begin{array}{c c c c} 1 & 2 & 3 & 4 \end{array} \\ \begin{array}{c} 1 \\ 2 \\ 3 \\ 4 \end{array} & \left( \begin{array}{c c c c} 0 & 1 & 2 & 3 \\ 1 & 0 & 1 & 2 \\ 2 & 3 & 0 & 1 \\ 1 & 2 & 3 & 0 \end{array} \right) \end{array}.$$

Since there may be delays due to ground service problems and/or weather in any of the locations, the shortest transit time from any city to any other may not necessarily be to fly directly, but may involve going indirectly via intermediate cities. So let's suppose the airline company wants to calculate the minimal time to reach city 4 from any of the other cities, given the above transit times between any pair of cities. We will first outline the dynamic programming procedure for solving all such problems, then return and use this procedure for this four-city problem.

Suppose we have a network of $N$ cities. For ease of notation, we label them $1, 2, \ldots, N$, and let $t_{ij}$ be the cost in going from city $i$ to city $j$. We want to find the minimal-cost path of going from, say, city 1 to city $N$. Using dynamic programming, we can easily solve this problem in the following way.

Define the quantity

$I_k^{(r)} =$ the minimal cost path from city $k$ to city $N$ having at
most $r$ intermediate stops, $r = 0, 1, 2, \ldots, N - 2$.

Clearly, if there are *no* intermediate stops on the path from city $k$ to city $N$, then the minimal cost is simply the cost of going directly to city $N$ from city $k$. By definition of the functions $I_k^{(r)}$, this means that $I_k^{(0)} = t_{kN}$.

Now suppose that we go from city $k$ to an intermediate city $j$. By definition, the cost of this transition is $t_{kj}$, and we have "spent" one intermediate stop in, say, a minimal-cost tour involving at most $r$ such stops. We are now faced with the problem of going from city $j$ to the termination city $N$ on a minimal-cost path with at most $r - 1$ intermediate stops. But, by definition, this minimal cost is just $I_j^{(r-1)}$. Putting these observations together, we are led to the inequality

$$I_k^{(r)} \leq t_{kj} + I_j^{(r-1)},$$

a relation that must hold for *any* city $j$ that we happen to choose as the first stop on our tour from city $k$. If we now choose city $j$ so as to minimize the right side of this inequality, we are led to the basic equation for the family of problems as

$$I_k^{(r)} = \min_{j \neq k} \left[ t_{kj} + I_j^{(r-1)} \right]. \qquad (¶)$$

211

Starting with tours having no intermediate stops ($r = 0$), we have the function $I_k^{(0)} = t_{kN}$, which is given to us by inspection. Using this function, we next set $r = 1$ and calculate $I_k^{(1)}$ from Eq. (¶). This function then allows us to compute the function $I_k^{(2)}$, and so on. Since the maximal number of intermediate stops on any tour is $N-2$, this process converges at a value of $r$ no greater than $N - 2$. At this point, we have not only the *optimal value function,* telling us the minimal cost for a tour from any city $k$ to city $N$, but also the *optimal policy function,* call it $j^*(k)$, telling us what city to go to next if we find ourselves at city $k$. And this holds for all cities $k = 1, 2, \ldots, N$. Now let's use these general ideas to solve the airline scheduling example.

First, we do the easy calculation of finding the optimal value function $I_k^{(0)}$ and the optimal policy function $j^*(k)$. Of course, $I_k^{(0)} = t_{k4}$, while the optimal policy when you are allowed no intermediate stops is to just go directly from whatever city you're in to city 4. These considerations immediately yield the results shown in Table 5.3.

Next, we compute $I_k^{(1)}$ and its associated optimal policy function for $k = 1, 2, 3$, and 4. Thus, when $k = 1$ we have

$$I_1^{(1)} = \min_{j\neq1} \left\{ t_{1j} + I_j^{(0)} \right\},$$

$$= \min \left\{ t_{12} + I_2^{(0)}, t_{13} + I_3^{(0)}, t_{14} + I_4^{(0)} \right\},$$

$$= \min\{1 + 2, 2 + 1, 3 + 0\},$$

$$= 3.$$

Thus, $I_1^{(1)} = 3$. Since this value is attained for any of the choices $j = 1, 2, 3$, we have $j^*(1) = 1, 2$ or 3. Carrying on, we obtain Table 5.4.

**Table 5.3** The optimal value and policy functions for $r = 0$.

| City $k$ | $I_k^{(0)}$ | $j^*(k)$ |
|---|---|---|
| 1 | 3 | 4 |
| 2 | 2 | 4 |
| 3 | 1 | 4 |
| 4 | 0 | 4 |

**Table 5.4** The optimal value
and policy functions for $r = 1$.

| City $k$ | $I_k^{(1)}$ | $j^*(k)$ |
|---|---|---|
| 1 | 3 | 1, 2, or 3 |
| 2 | 2 | 3 or 4 |
| 3 | 1 | 4 |
| 4 | 0 | 4 |

Comparing Tables 5.3 and 5.4, we see that $I_k^{(0)} = I_k^{(1)}$. Therefore, the process has converged, which means the optimal cost and policy associated with $r = 0$ (that is, no intermediate stops) is optimal for this especially primitive network of cities. In other words, for this network the minimal-cost path in going from any city to city 4 is what we might call the "Monopoly solution": go directly to city 4 without passing through any of the other cities along the way. Of course, in more complicated networks involving dozens, or even hundreds, of cities, we can expect the minimal-cost in going from one city to another to involve lots of intermediate stops. In fact, a good illustration of this point is the way airline ticket costs are set up nowadays, since the cheapest tickets usually involve transferring at an airline "hub" instead of going directly to your destination. Now let's employ the same general imbedding idea to a slightly more realistic version of the resource allocation problem discussed early in the chapter.

## Getting the Most for Your Money

Suppose you are an investor with a certain amount of money at your disposal. There are $N$ different kinds of investments available to you, ranging from very risky, high-yield speculations like Moscow real estate to moderate-risk, moderate-yield blue chip stocks, on to no-risk, low-yield government bonds. Assume that you have a total of $X$ dollars of investing capital at your disposal, and that there is a known expected rate of return from each of the $N$ possible investments. Then your problem is how to parcel out your capital among the $N$ possible investments so as to maximize the annual return. Here's a simple example of this kind of problem that again illustrates dynamic programming in action.

213

Let there be $N = 3$ different investments: stocks, bonds, and commodities. Furthermore, assume that an investment of $y$ dollars in stocks yields an annual rate of return of $g_1(y)$ dollars. Similarly, an investment of $y$ dollars in bonds produces an annual yield of $g_2(y)$, while the same amount invested in commodities leads to a return of $g_3(y)$ per year. If our allocation is $x_S$ dollars to stocks, $x_B$ dollars to bonds and $x_C$ dollars to commodities, the problem becomes that of maximizing the total return $g_1(x_S) + g_2(x_B) + g_3(x_C)$, subject to the constraint that we can't invest more than we have. Mathematically, this translates into the inequality $x_S + x_B + x_C \leq X$. Here's how to solve this problem via dynamic programming.

First of all, we note that if stocks are the only investment available (the case $N = 1$), then the problem is easy. We simply maximize $g_1(x_S)$ over all levels of investment $x_S$ such that $0 \leq x_S \leq X$. Next we observe that the solution of the investment problem depends on both $N$, the number of investments available, and on $X$, the total capital. So let's define the function

$I_3(X) =$ the maximum possible return when there are three possible investments and $X$ dollars of capital is available.

Now imagine that we allocate an amount $x_C$ to investment $N = 3$, commodities. Two things happen as a result of this allocation: (a) the available capital is reduced to $X - x_C$, and (b) the number of investment opportunities remaining to which this capital can be allocated decreases from $N = 3$ to $N = 2$. Moreover, whatever the allocation $x_C$ made to commodities (investment opportunity 3, the remaining capital must be used to maximize the return over the two remaining investment modalities, stocks and bonds. This is a particular instance of what Bellman christened the "Principle of Optimality." Roughly speaking, it says that "Any part of an optimal policy must be optimal," an almost self-evident truth—but one that can be used to great practical advantage.

We can invoke the Principle of Optimality to express the above argument in symbolic terms as $I_3(X) \geq g_3(x_C) + I_2(X - x_C)$, which follows immediately from the definition of the function $I_3(X)$ as the *greatest* possible return for an three-investment process with total available capital $X$. This inequality must hold for *any* allocation $x_C$ to commodities. So we obviously want to choose $x_C$ to maximize the right-hand side of the above inequality. This leads to the following equation relating an allocation processes of three investment opportunities to one in which there are two investments available:

$$I_3(X) = \max_{0 \leq x_C \leq X} [g_3(x_C) + I_2(X - x_C)], \quad N > 1. \qquad (\maltese)$$

As noted above, we can easily compute $I_1(X)$ as

$$I_1(X) = \max_{0 \leq x_S \leq X} g_1(x_S). \qquad (\maltese\bullet\maltese)$$

Eqs. $(\maltese)$ and $(\maltese\bullet\maltese)$ can be used to solve the original investment problem in the following way. First, use Eq. $(\maltese\bullet\maltese)$ to determine the function $I_1(X)$. This involves solving an optimization problem for each possible value of the capital $X$. Making use of the function $I_1(X)$, we now determine the function $I_2(X)$ using Eq. $(\maltese)$. Knowing the function $I_2(X)$, we can now determine the functions $I_3(X)$. What this procedure accomplishes is to take the original problem involving a *particular* amount of capital and a *fixed* number of investment opportunities, and imbed it within a family of problems in which both the available capital and the number of possible investments can vary. Eqs. $(\maltese)$ and $(\maltese\bullet\maltese)$ provide a scheme by which we can solve this family of problems, after which we may then just pluck out the particular one that's of interest.

The above examples of allocation processes and routing problems only begin to suggest the kinds of optimization problems that dynamic programming allows us to attack. Optimal control processes, like getting a satellite into a particular orbit with minimal energy, problems with random influences such as investments with risky payoffs, questions involving unusual criteria like maxmin of game theory (see Chapter 1) are all within the purview of the theory. So why don't we use dynamic programming as a universal panacea for every type of optimization problem? The answer lies with what Bellman has termed the "curse of dimensionality." This appellation refers to the exponential growth in computer resources needed to solve a problem by dynamic programming as the problem size increases. The material cited in the bibliography gives an ample account of this barrier to our ever reaching optimization nirvana.

# Bibliography

*Note:* Rather than being an exhaustive account of each source I consulted in preparing this book, as would be right and correct in a research monograph or scholarly treatise, the following list of references is intended to be more of a set of pointers to the literature. Thus, it constitutes a somewhat eclectic collection of papers, books and articles that the reader may want to examine to get more details on the topics considered here, as well as to investigate many of the themes that were not considered in this book due to either space constraints or a level of technical background required going beyond the book's target reader. The material cited follows more or less sequentially the topics as they appear in the body of the text.

## Chapter 1 (Game Theory)

Williams, J. *The Compleat Strategyst.* New York: McGraw-Hill, 1954. A purely expository work that's accessible to just about anyone who can count, add, and subtract, and that gives a instructive and entertaining account of the subject by means of many illustrative and amusing examples.

Thomas, L. C. *Games, Theory and Applications.* Chichester, England: Ellis Horwood, 1986. A first-rate, undergraduate-graduate text on the theory of games with many examples, including the Fighter-Bomber Game used in the text. Excellent discussion of the use of linear programming for calculating the optimal mixed strategies for general two-person, zero-sum games. Also includes a good account of $n$-person, cooperative games.

Von Neumann, J. and O. Morgenstern. *Theory of Games and Economic Behavior.* Princeton, NJ: Princeton University Press, 1944. The classic work from which the entire field emerged almost full-grown.

Colman, A. *Game Theory and Experimental Games.* Oxford: Pergamon Press, 1982. Contains a wealth of examples of game theory in action, especially in the social and behavioral sciences.

Jones, A. *Game Theory.* Chichester, England: Ellis Horwood, 1980. An outstanding mathematically-oriented introduction to the subject, which includes the Concord Arsenal Game discussed in the text.

Von Neumann, J. "Zur Theorie der Gesellschaftsspiele." *Math. Annalen,* 100 (1928), 295–320. Von Neumann's original proof of the Minimax Theorem.

Brams, S. *Game Theory and Politics.* New York: Free Press, 1975. Discusses a wealth of examples involving military and international relations situations in which game-theoretic ideas enter heavily, including the Cuban Missile Crisis Game of the text.

*Game Theory and Related Approaches to Social Behavior,* M. Shubik, ed., New York: Wiley, 1964. An excellent summary of much of the early work in game theory, especially as it relates to conflict situations in social sciences and economics.

Shubik, M. *Game Theory in the Social Sciences.* Cambridge, MA: MIT Press, 1982. A first-rate summary of the various branches of game theory as applied in politics, economics, and the social arena, in general.

Rapoport, A. and A. Chammah. *Prisoner's Dilemma: A Study in Conflict and Cooperation.* Ann Arbor, MI: University of Michigan Press, 1965. A full and complete treatment of the Prisoner's Dilemma in all its many guises.

Maynard Smith, J. *Evolution and the Theory of Games.* Cambridge: Cambridge University Press, 1982. The definitive work on ESS and its applications in population ecology.

Axelrod, R. *The Evolution of Cooperation.* New York: Basic Books, 1984. A first-hand account of the fascinating computer experiments with strategies for playing the iterated Prisoner's Dilemma Game.

# Chapter 2 (Topology)

Shashkin, Yu. *Fixed Points.* Providence, RI: American Mathematical Society, 1991. Translation of a Russian introduction to the topic written for bright high-school students. Lots of geometrical insight and many challenging and thought-provoking exercises.

Stewart, I. *Concepts of Modern Mathematics.* London: Penguin, 1975 (Paperback edition: New York: Dover, 1995). This little paperback volume contains a wealth of information about modern mathematics—including four chapters on topology—written in the author's inimitable style. Not too difficult for the layperson, not too superficial for the expert. Highly recommened.

Jacobs, K. *Invitation to Mathematics.* Princeton, NJ: Princeton University Press, 1992. A first-rate educated layperson's introduction to many of the central themes of modern mathematics. Somewhat more technical than the preceding volume, but still very readable. Contains an excellent chapter on topology.

Hansen, V. *Geometry in Nature.* Wellesley, MA: A. K. Peters, 1993. An extraordinarily good introduction to geometry and its uses in studying the physical world. Contains outstanding chapters on both the topology of surfaces as well as the topology of catastrophes.

Alexandroff, P. *Elementary Concepts of Topology.* New York: Dover, 1961. Gentle, intuitive 60-page introduction to topology written at a level accessible to undergraduates.

Arrow, K. and F. Hahn. *General Competitive Analysis.* Amsterdam: North-Holland, 1971. Classic mathematical account of general equilibrium analysis.

Debreu, G. "Four Aspects of the Mathematical Theory of Economic Equilibria." *Proc. Int. Congress of Mathematicians.* Vancouver, BC: Canadian Mathematical Congress, 1974, 65–77. Good, but fairly technical, survey of the mathematics of general equilibrium theory.

Smart, D. R. *Fixed Point Theorems.* Cambridge: Cambridge University Press, 1974. Summary of most of the known mathematical results about fixed points. Quite technical. For professionals only.

Brouwer, L. E. J. "Über Abbildung von Mannigfaltigkeiten." *Math. Annalen,* 71 (1910), 97–115. Brouwer's original article on the mapping of manifolds, in which he presents his fixed point theorem.

Scarf, H. "Fixed-Point Theorems and Economic Analysis." *American Scientist,* 71 (May-June 1973), 289–296. Introductory account of how fixed points arise in economic analysis, together with a good discussion of the simplicial approximation method.

Scarf, H. with the collaboration of T. Hansen. *The Computation of Economic Equilibria.* New Haven, CT: Yale University Press, 1973. Detailed mathematical account of how to use the simplicial approximation method to find fixed points in economics.

Keener, J. "The Perron-Frobenius Theorem and the Ranking of Football Teams." *SIAM Review,* 35 (March 1993), 80–93. Mathematical account of four different methods to rank teams in uneven pairwise competition.

# Chapter 3 (Singularity Theory)

Stewart, I. and T. Poston. *Catastrophe Theory.* London: Pitman, 1978. An excellent account of the theory of singularities and catastrophes with many examples from physics, engineering, and biology. Moderately technical, but not just for mathematicians. Outstanding bibliography listing just about everything in print up to the book's publication date.

Lu, Y. C. *Singularity Theory and an Introduction to Catastrophe Theory.* New York: Springer, 1976. Nice mathematical introduction to the ideas presented in this chapter. Fairly mathematical.

Deakin, M. "An Elementary Approach to Catastrophe Theory." *Bull. Math. Biol.,* 40 (1978), 429–450. Account of the ideas underlying singularity theory and catastrophe theory using nothing more than first-year calculus and the idea of a power series expansion.

Woodcock, T. and M. Davis. *Catastrophe Theory.* New York: Dutton, 1978. Layperson's treatment of catastrophe theory. Completely nonmathematical. Excellent discussion of applications of the theory, as well as a detailed account of the "catastrophe controversy."

Woodcock, A. E. R., and T. Poston, *A Geometrical Study of the Elementary Catastrophes.* Springer Lecture Notes in Mathematics, Vol. 373. Berlin: Springer, 1974. Beautiful computer drawings of the sections of the various higher-dimensional bifurcation surfaces.

Saunders, P. *An Introduction to Catastrophe Theory.* Cambridge: Cambridge University Press, 1980. Gentle, compact introduction to catastrophe theory for the student. Develops an amazing amount of material with a minimal amount of mathematics.

Thom, R. *Structural Stability and Morphogenesis.* Reading, MA: W. A. Benjamin Co., 1975. This is the book that started it all. A remarkable combination of mathematical arguments, biological speculation, and philosophical insights. Not to be missed.

Stewart, I. "Applications of Catastrophe Theory to the Physical Sciences." *Physica D*, 20 (1981), 245–305. Survey article outlining the uses of catastrophe theory in physics and engineering.

Zeeman, E. C. *Catastrophe Theory: Selected Papers, 1972–1977.* Reading, MA: Addison-Wesley, 1977. Collection of articles showing catastrophe theory in action. The section on applications in the social sciences contains many of the papers that set off the critics of catastrophe theory.

Baillieul, J. and C. Byrnes. "A Geometric Problem in Electric Energy Systems," in *International Symposium on the Mathematical Theory of Networks.* Vol. 4. Hollywood, CA: Western Periodicals, 1981. An interesting use of singularity theory in the area of electrical energy production.

Thom, R. "Topological Models in Biology." *Topology,* 8 (1969), 313–335. Detailed account of Thom's ideas about biological form that motivated him to develop the mathematical theory of catastrophes.

Sussman, H. J. "Catastrophe Theory: A Preliminary Critical Study." *Proc. Biennial Mtg. Phil. Sci. Assn.,* 1976. This paper fired the opening salvo in the case against catastrophe theory.

Arnold, V. I. *Singularity Theory.* Cambridge: Cambridge University Press, 1981. Fairly technical collection of articles detailing the mathematics behind singularity theory.

Gilmore, R. *Catastrophe Theory for Scientists and Engineers.* New York: Wiley, 1981. Very good engineering discussion of both the theory and applications of catastrophe theory. Moderately mathematical—at the kind of informal level congenial to engineers.

Paulos, John Allen. *Mathematics and Humor.* Chicago: University of Chicago Press, 1980. Introduction for the general reader to the analysis of humor using mathematical tools—logic, Gödel's Theorem, and catastrophe theory. Very entertaining. Illustrates Wittgenstein's famous remark that a serious work of philosophy could be written that consisted of nothing but jokes.

## Chapter 4 (Theory of Computation)

Rucker, R. *Mind Tools.* Boston: Houghton-Mifflin, 1987. A fine introductory account of modern mathematical thinking, including an excellent discussion of Turing machines, computation, and their relation to the mind.

Davis, M. *Computability and Unsolvability.* New York: McGraw-Hill, 1958 (reprint edition: New York: Dover, 1982). Technical introduction to the notion of computability. Also includes a reprint of Davis's well-known article on the unsolvability of Hilbert's 10th Problem.

Epstein, R. and W. Carnielli. *Computability: Computable Functions, Logic and the Foundations of Mathematics.* Pacific Grove, CA: Wadsworth & Brooks/Cole, 1989. One of the most readable and intuitive introductions to computability theory that I know of. Notable for its many excerpts from the work of Hilbert, Gödel, and others on the philosophy of mathematics and its relationship to computability theory.

R. Herken, ed. *The Universal Turing Machine: A Half-Century Survey.* Oxford: Oxford University Press, 1988. A stimulating collection of essays reviewing current knowledge about Turing machines and their many implications and ramifications in other areas.

Jones, J. "Recursive Undecidability: An Exposition." *Amer. Math. Monthly,* 81 (1974), 724–738. An illuminating survey article on undecidability and computation, which contains a proof of the uncomputability of a winning strategy for the Turing Machine Game.

Hofstadter, D. *Gödel, Escher, Bach: An Eternal Golden Braid.* New York: Basic Books, 1979. A fine introduction to the notions of formal systems, Gödel's Theorem, and artificial intelligence. Includes an account of the ★-⚔-☀-system used in the text.

Nagel, E. and J. R. Newman. *Gödel's Proof.* New York: New York University Press, 1958. The first account of Gödel's Theorem written expressly for the general reader—and still one of the best.

Chaitin, G. *Information, Randomness, and Incompleteness.* 2d ed. Singapore: World Scientific, 1990. A reprint collection telling the complete story of Chaitin's independent discovery of algorithmic complexity and its connection with randomness.

Garey, M. and D. Johnson. *Computers and Intractability.* San Francisco: Freeman, 1979. The definitive general reference on computational tractability, $NP$-completeness, and other matters of this ilk.

Blum, L., M. Shub, and S. Smale. "On a Theory of Computation and Complexity over the Real Numbers: $NP$-Completeness, Recursive Functions and Universal Machines." *Bull. Amer. Math. Soc.,* 21 (1989), 1–46. Definitive mathematical account of the Blum–Shub–Smale machine model of computation.

Chaitin, G. "The Limits of Mathematics." IBM Research Report RC-19646, July 1994. A hands-on, computationally oriented exposition of algorithmic information theory. Contains LISP code for actually calculating many of the complexity bounds that appear in the theory. (This document can be downloaded from the Internet by sending an email to: chao-dyn@xyz.lanl.gov with Subject:get 9407003.)

A. R. Anderson, ed. *Minds and Machines.* Englewood Cliffs, NJ: Prentice Hall, 1964. Reprints of articles on the philosophical aspects of artficial intelligence. Contains Turing's famous 1950 article on artificial intelligence, as well as Lucas's (in)famous 1961 Gödelian-based counterattack.

Bennett, C. "On Random and Hard-to-Describe Numbers." IBM Research Report RC-7483 (#32272), IBM Research Laboratories, Yorktown Heights, NY, May 23, 1979. Informal discussion of the properties of Chaitin's number $\Omega$, together with its implications for the decidability of conjectures in mathematics.

Shub, M. "Mysteries of Mathematics and Computation." *Math. Intelligencer,* 16 (1994), 10–15. Introductory account of the relationship between the theory and the practice of computation. Includes an excellent summary of the results obtained (through 1993) using the BBS model of computation.

# Chapter 5 (Optimization Theory)

E. Lawler, et al, eds. *The Traveling Salesman Problem.* New York: Wiley, 1985. Good overview of the computational state-of-the-art for solving the Traveling Salesman Problem.

Darst, R. *Introduction to Linear Programming.* New York: Dekker, 1991. Very good introduction to linear programming and its many extensions— networks flows, quadratic programming, integer programming, and so forth. Includes lots of worked exercises and examples.

Dantzig, G. *Linear Programming and Extensions.* Princeton, NJ: Princeton University Press, 1963. All-time classic account by the founder of linear programming. Almost encyclopedic, containing absolutely everything known about the topic at the time.

Vadja, S. *Theory of Games and Linear Programming.* London: Methuen, 1956. Concise, easy-to-understand account of LP and its relation to two-person, zero-sum games of strategy.

Ford, L. R. and D. R. Fulkerson. *Flows in Networks.* Princeton, NJ: Princeton University Press, 1962. Full-blown account of the Min Cut-Max Flow Theorem and its many applications.

Ahuja, R., T. Magnanth, and J. Orlin. "Some Recent Advances in Networks Flows." *SIAM Review,* 33 (1991), 179–219. Up-to-date account of developments for solving network flow problems. Emphasizes algorithmic aspects of flow problems.

Borgwardt, K. H. *The Simplex Method.* Berlin: Springer, 1987. A detailed account of why the Simplex Method is so efficient in practice, despite the fact that its worst-case analysis shows it to be computationally hard.

Arbel, A. *Exploring Interior-Point Linear Programming.* Cambridge, MA: MIT Press, 1993. In recent years, alternatives to the Simplex Method have been developed that use projections out from the interior of the feasible region to find the optimal point on the boundary. This book gives an excellent introduction to these methods, which are often considerably faster than the Simplex Method. Includes a computer diskette with programs for the interior-point procedures.

Dolan, A. and J. Aldous. *Networks and Algorithms.* Chichester, England: Wiley, 1993. Excellent introductory text on network flow problems based on the Open University course on graphs, networks, and designs. Highly recommended.

Bellman, R. E. and S. E. Dreyfus. *Applied Dynamic Programming.* Princeton, NJ: Princeton University Press, 1962. Classic work on dynamic programming. Emphasises applications in control theory and operations research.

Larson, R. and J. Casti. *Principles of Dynamic Programming–Parts I and II.* New York: Dekker, 1978, 1982. Introduction to dynamic programming. Treatment is almost exclusively by hundreds of worked examples, emphasizing the numerical aspects of the approach.

Saaty, T. *Nonlinear Mathematics.* New York: McGraw-Hill, 1964. Outstanding account of all types of nonlinear problems and methods for their solution. Chapter on nonlinear programming is especially good, as it's filled with methods and worked examples circa mid-1960s.

Künzi, H., H. Tzschach, and C. Zehnder. *Numerical Methods of Mathematical Optimization.* New York: Academic Press, 1968. Overview of many of the most popular computational methods for solving nonlinear optimzation problems. Includes Fortran and Algol program codes for most of the procedures.

Bellman, R. E. "Mathematical Aspects of Scheduling Theory." *SIAM J. Appl. Math.,* 4 (1956), 168–205. Excellent introduction to the use of dynamic programming for solving problems assembly-line scheduling, network flows, inventory control and lots more. Includes a discussion of the comparisons of dynamic programming methods with LP, pointing out the relative advantages of each approach.

# Index

$\mathcal{G}$ (directed graph or digraph), 200
$\Gamma$ (gamma), 179
$\Omega$ (omega), 171–74
$\Phi$ (phi), 179–80

Affine geometry, 51
Algebra, xi, 49
Algorithm, 137–38, 179
    complexity, 171
    Euclidean, 138–40
    gradient method, 208–9
    Simplex Method, 190
Allocation processes, 213–15
Alphabet (formal mathematical system), 154
Analysis (mathematics), 49
Analytic geometry, 51
A/P (Area per person), 205
Arcs, 200
Area per person (A/P), 205
Aristotle, 153
Arithmetic, theory of, 157
    Gödel's theorem, 158–65
Armstrong, Neil, 78
Arnold, V. I., 113

Arrow, Kenneth, 64
Assignment problem, 177
Atiyah-Singer Index Theorem, xii
Axelrod, Robert, 36–38
Axiom (formal mathematical system), 153–54
Axiomatic system, 153

Balasko, Yves, 64
Basic solution, 192
Basis, 192
Battle of the Bismarck Sea, 3–8, 16
Battle of the Sexes, 9, 32–33
Bellman, Richard, 210
    Principle of Optimality, 214–15
Benacerraf, Paul, 168
Bennett, Charles, 173
Berry, Michael, 113
Bifurcation diagram, 119
Bifurcation point, 119
Bifurcations, 115, 116, 119–23
Bin-packing problem, 178
Blum, Lenore, 179
Blum-Shub-Smale (BSS) model, 179–80

227